AF836298

DISTINCTION
JOURNAL OF FORM

Volume 1 Issue 2

Winter 2025

The Spencer-Brown Society

ISBN: 978-1-84890-503-0

Published by College Publications, London.

In Memoriam

James M Flagg
1949 - 2025
Scholar

Editorial Introduction

Distinction: Journal of Form provides a channel for the publication of work by investigators who draw on the work of George Spencer-Brown, and develop it, often in diverse and surprising ways. The Society is steered by a committee, elected at the 2024 AGM, comprising Graham Ellsbury (Chair), Florian Grote (Secretary), Leon Conrad (Treasurer), Divyamaan Sahoo (Publications Director) and Randolph Dible (Communications Director). Fred Cummins and Philip Franses have provided significant editorial and compositing work on this Journal and additional administrative support.

The Society is embarking on a programme to develop a substantial website, with Leon taking on the new role of Website Content Director. This will leave the position of Treasurer vacant. A provisional Treasurer will be appointed by the committee as an interim measure, and a Treasurer will be formally elected by the membership at the AGM in August 2026.

The Society welcomes applications from UK-resident (as required by the Society's bankers) Society members, for this voluntary role.

The Society extends its warmest thanks to Leon for the efficient and diligent manner in which he has discharged the role of Treasurer of the Society.

Distinction: Journal of Form, Volume 1 Issue 2, Winter 2025

In 'The Sense of Sentences', Leon Conrad uses Spencer-Brown's work to demonstrate the need to reexamine the way we teach language. He argues that sentence types differ in terms of the qualities of the relationship between the knower and the known which give rise to them. Using expressions outlined in *Laws of Form*, he demonstrates that the widely recognised set of four sentence types (statement, question, order, exclamation) should be extended to seven. While the paper focuses on the English language, readers are invited to test out its wider applicability by applying its findings to other languages themselves.

Nathaniel Hellerstein's paper, 'Diamond Logic and How to Count to Two' considers several well-known results in the foundations of mathematics: the liar paradox, Russell's paradox, Cantor's paradox and Gödel's first incompleteness theorem. In doing so, Hellerstein employs four truth-values: the standard Boolean values of true and false, and two oscillating values, arranged in the form of a diamond lattice. These values are then used to investigate the properties of modulators and rotors. The ensuing enquiry summarizes, with some modification, parts of Chapters 1, 2, 3, 7, and 11, and most of Chapter 12, from his work:

> *Diamond, A Paradox Logic*, 2nd Ed,
> N. S. Hellerstein,
> Copyright © 2010 by World Scientific Publishing Co. Pte. Ltd.
> ISBN-13: 978-981-4287-13-5; ISBN-10: 981-4287-13-X.

The editors particularly wish to thank World Scientific Publishing for their kind permission to publish Hellerstein's paper here.

The article 'Flourishing of the Form—Co-arising of Possibility Fields' by Walter Tydecks follows Husserl in asking: How does theory contend with elements—like imaginary numbers—that have no simple definition in themselves? The "flourishing of the form" describes the way that a form must ascend through the theory in order to achieve an identity that is not there at the beginning. The theory must include how the thing in itself, kinds of givenness, ways of taking and hearing, relate to the movement that accommodates all these aspects in a picture able to settle back into the terms of the original conceptual foundation. It is here that Spencer-Brown excels, in precisely accomplishing this task in *Laws of Form*. Husserl's test of true theory flourishes under the mathematical laws of Spencer-Brown, arising into form.

The article 'Distinct but Not Exclusive Categories: Introducing the n-category-lemma for second order observation' by Steven Watson explores how Spencer-Brown's arithmetic is more suited to an extended four-valued logic (true, false, true and false, neither true nor false). From the standpoint of this extended logic, the re-entry of form in Spencer-Brown, instead of being taken as a separate implication of mark/unmark, completes the foundation of distinction as that which is both marked and unmarked. There is a natural inclusion of the re-entry into the notion of mark. It is also shown how this approach applies to Luhmann's general theory of autopoiesis, changing the way we deal with categories by opening a space of interactivity and mutual influence. The multi-dimensional logical states also provide a way to help Large Language Model generative AI delve into logical relations. The article showcases how and why *Laws of Form* is so revolutionary in the way that it deals with re-entering marks.

In the article 'C. S. Peirce Meets George Spencer-Brown: Seeking semiotic self-evidence for logic,' Jonathan J. Mize invites the reader to "dig even deeper than the distinction, seek even simpler, more primitive forms of expression that can encapsulate logic and mathematics." The author asks whether, on first meeting, a student would immediately intuit what Pierce's or Spencer-Brown's diagrams were about. Or if not, is there a deeper bedrock that can reveal this inner content more spontaneously? To achieve this representational self-evidence, the article makes use of Hans Freudenthal's "Lincos" language devised to be understood by extraterrestrial life (and first-time readers). This is extended to a pictorial interface that employs a domino like simplicity of form to present logical propositions. The approach offers an aesthetic and intuitive simplicity for grasping (even at the first time of reading) a pattern across the given examples. It also seems to erase the boundaries between signs, operators, symbols, and statements, fulfilling its stated aim of becoming a foundational semiotics.

In 'Towards A New Monadology', Dirk Baecker proposes a reinterpretation of Leibniz, viewing monads as the prerequisite for making distinction and indication a possibility. Baecker argues that monads *are* communication and, instead of through interacting with God, find other monads as observations within themselves. In doing so, monads operate simultaneously as the forms of communication and the medium for the communication in all other monads they observe.

William Bricken's 'The Numeric Forms of James Algebra' details a mathematical system, due to Jeffrey James, that fundamentally reconstructs a complete algebra of numbers (including integer, fractional, real, and imaginary expressions) using the core idea of "distinction" in Spencer-Brown's *Laws of Form* with a new interpretation. Using two types of containers ( ) and [ ], James algebra rests on three pattern axioms, and introduces concepts such as that of a non-numeric unit, a generic inverse, and a base-free approach to exponents. The round boundary ( ) functions as the archetypal numeric unit "one" as well as an exponentiation operator, depending on the context. The square boundary [ ] is a non-numeric unit, interpreted as a logarithm. This system offers a unique perspective on base-free exponents and logarithms, where # is an "arbitrary base". "Frames" are specific containment patterns where an outer round-bracket contains at least one square-bracket, e.g., `(A[B])`. There are three pattern axioms: Inversion, Arrangement, and Reflection. Inversion defines the two boundaries as mutual annihilators, such that the pair (`A[A]`) is void-equivalent. Arrangement specifies how multiple boundaries interact and is the source of all structural complexity, acting as the system's distribution rule. Reflection introduces the constant unit J (interpreted as $\log_{\#} -1$) as a generalized inverse indicator, reversing constructive operations without requiring separate inverse functions. This model provides a uniform computational structure where all number types (natural, integer, complex, etc.) and all arithmetic operations (including subtraction, division, and roots) can be expressed.

In 'Primes Between Squares: Commentary on Appendix 8 of *Laws Of Form*,' J. M. Flagg, Louis H. Kauffman and Divyamaan Sahoo provide annotations to Spencer-Brown's work on Legendre's conjecture that there exist at least two prime numbers between consecutive squares n^2 and $(n+1)^2$. The commentary is a guided tour of Appendix 8 of *Laws of Form*. The foundation of Spencer-Brown's approach to the Sieve of Eratosthenes is demonstrated, and so is the "prime paradigm", which is a repeating, four-fold symmetric pattern of composites generated by a set of primes. Defining "sect" as a consecutive sequence of composite numbers and "segment" as the stretch of numbers between squares, the question is: What is the maximum sect that can appear in a square segment? If the maximum sect length is less than the square segment length, then there exists at least one prime in that segment. Two conjectural lemmas (Lemma 3 and 4) describe the form of maximal odd sects (sequence of odd composites) as either "quadrantic" or "reflexive", and observe that the "quadrantic" sect length is always less than the "reflexive" one. Using these two Lemmas, a proof of Theorem 1 is provided. Theorem 1 is generalized to a family of conjectures that the next prime after p is at most $\frac{1}{k}\sqrt{p}$ away, where k starts at $\frac{1}{2}$ (Theorem 1A) and increases to a "ratchet" value from which it does not drop at primes $p = 7, 113, 1327, \ldots$ (Theorem 1B), culminating in a Corollary that the previous prime is in $(p - \sqrt{p}, p)$ and the next prime is in $(p, p + \sqrt{p})$ for all $p \geq 131$, and Theorem 2 that there exist two primes between squares. Spencer-Brown's broader observation is that the number of primes between squares is asymptotic to $\frac{n}{\log n}$. The commentary closes with Spencer-Brown's proof of Theorem 1 using "elementals", connecting the sect formation algorithm provided by Spencer-Brown in Lemma 3 with the modulator E4 in Chapter 11 of *Laws of Form*. Three appendices are included, which contain sharper conjectures provided by Spencer-Brown, how they relate to other well-known conjectures in number theory, and an illustration of Lemma 3's sole exception at 23.

The editors would like to extend their warm thanks to the reviewers for their work in reviewing the papers in this issue.

Finally, it is intended that future issues of *Distinction* will feature a letters page. Readers wishing to express an opinion, or comment on any matter concerning George Spencer-Brown's life and works, the LoF conferences, or this journal are invited to write to the editors for possible inclusion in *Distinction: Journal of Form.*

The Editors

Divyamaan Sahoo
Fred Cummins
Leon Conrad
Graham Ellsbury
Florian Grote
Randolph Dible
Philip Franses

Laws of Form Conference, 2026

We are delighted to announce that LoF26, the fourth Laws of Form conference, will be held from Monday 10 August 2026 to Friday 14 August 2026 at:

The Faculty of Education
University of Cambridge
184 Hills Road
Cambridge CB2 8PQ
United Kingdom

There is no charge for attending the conference, and we are greatly indebted to the Faculty of Education, and to Cambridge University, for providing the conference venue free of charge.

In addition to the conference, it is intended that there will be additional activities, including:

- Punting on the River Cam
- Evensong at King's College Chapel
- A visit to The Eagle pub, where in 1953, Francis Crick announced that he and James Watson had discovered "the secret of life".

A programme detailing these and other activities, will be announced in due course on the Society's website at lof50.com

We look forward to welcoming you to LoF26 in Cambridge, August 10–14, 2026!

The Sense of Sentences

Leon Conrad
The Next Society Institute
leon@traditionaltutor.co.uk

Abstract

How many sentence types are there? What does George Spencer-Brown's *Laws of Form* (2011) offer in terms of helping us reach an answer? This paper focuses on the intentions which give rise to different sentence types in the English language. These intentions are mapped to expressions outlined by Spencer-Brown. The paper concludes that the approach reveals hitherto unappreciated subtleties at the very ground of language and puts forward suggestions for further work in the application of Spencer-Brown's work to the study of language.

In the current UK National Curriculum for English (HM Government, 2013, 15), a sentence is defined as 'a group of words grammatically connected to each other but not to any words outside the sentence.' The definition continues: 'The form of a sentence's main clause shows whether it is being used as a statement, a question, a command or an exclamation.' In the US, since 2010, from Grade One, students are expected to 'produce and expand complete simple and compound … sentences' (Common Core 2010, 26). And while two sentence structures (simple and compound) are referred to, the sentence types listed are the same as in the UK: declarative (statements), interrogative (questions), imperative (orders or commands) and exclamatory sentences (exclamations). In Australia, the curriculum currently requires students to recognise the 'different types of punctuation' associated with sentences 'that make statements, ask questions, express emotion or give commands'. (Australian Curriculum 2010, ACELA1449) While the same four sentence types (declarative, interrogative, imperative, and exclamatory) are listed, not only is the focus of the curriculum limited to written forms of language, it also asks readers to accept that it is the sentences that state, ask, express, and give, rather than the speakers or writers who produce those sentences that do so. While three national examples are given here, they exemplify a standard grouping of four sentence types that can be found in other international curricula and text books.

What if there is more to sentences than students are being taught at schools? And what does George Spencer-Brown's *Laws of Form* (2011) have to do with this?

In *Laws of Form*, Spencer-Brown outlines a calculus which shows that the basic forms underlying, among other things, language as a form, all emerge from two basic ideas: the idea of distinction and the idea of indication (2011, 1; see also Conrad 2020–23 for a series of videos on YouTube which goes through a close reading of the main body of Spencer-Brown's text). If we see the mark of distinction as being fixed and the act of indication as being dynamic, then these two basic ideas can be traced in sounds, the basic building blocks of language. Whether we see vowels as fixed or dynamic and consonants as dynamic or fixed, we observe this play of the fixed and the dynamic in the two types of sounds which we use to form words. Both types of sounds are generated through time. Both types of sounds have dynamic and fixed aspects. This play of fixed and dynamic is also there in the parts of speech we use in English which can be grouped into fixed and volatile types of words, an approach I take when teaching the parts of speech. I see the noun family as being comprised of words with 'fixed' qualities: either these are words which are linked directly to the fixed quality in the case of nouns, or they are words which connect to nouns in some way, e.g. adjectives; pronouns; articles; and prepositions, which are noun-joining. I see the verb family as being comprised of words with 'volatile' qualities: either these have a direct 'volatile' quality, e.g. verbs, or they are words which connect to verbs in some way, e.g. adverbs and conjunctions. The latter are verb-joining. While adverbs also modify adjectives and other adverbs in English, as modifiers of modifiers, they feature a dynamic rather than a static quality. The final part of speech, the interjection, straddles both groups, as it embodies both fixed and volatile qualities. Take 'Huh?!' or 'Ah!' as examples. You cannot have an 'Ah!' without there being someone to express the sentiment, and the sentiment cannot be expressed without the action taking place through time.

We cannot have sentences unless we have words. We cannot have words unless we have sounds. At each level, we see basic fixed and dynamic qualities at work. What about sentences? Can we map the fixed and dynamic qualities to sentence types? And if so, how?

If we define a sentence as a complete thought, then 'Jack and Jill went up the hill', being a complete thought, is an example of a sentence. However, while 'Jack and Jill' or 'only two' are meaningful groups of words, neither group forms a complete thought.

If I were to say "Jack and Jill" out of context, you would probably respond, "Yes ... What about them?" The same applies to 'went up the hill'. Hear these out of context, and you would probably ask, "Who or what went up the hill?" A complete thought needs a substantive (a substance word, a noun or pronoun) and a verb. This is one of the meanings of the Greek logos (Plato: Cratylus, 262c). The noun or pronoun will be static, the verb will be dynamic. Either can be stated or implied. The verb will generally denote an action that takes place through time. A sentence, in other words, is a *logos*, a coupling of the static and the dynamic. In spoken form, it is easy to distinguish between 'Jack and Jill went up the hill' and 'Jack and Jill went up the hill!' The difference is distinguishable through the quality of the speech act which originates in a difference in the qualities of the thoughts which give rise to each type of sentence. The structural differences which feature in the interrogative 'Did Jack and Jill go up the hill?' and the imperative 'Jack and Jill, go up the hill' highlight the difference in intention. There will also be intonational differences in the speech acts. These ultimately link to further differences in the qualities of the thoughts which give rise to these sentence types. Thus, sentence types differ in terms of the qualities of the relationship between the knower and the known which give rise to them. This is nothing new. It goes back to Aristotle, maybe even before him. Aristotle's *Peri Hermeneias* (*De Interpretatione*) was transmitted to the West by Boethius (Arens 1984, 159—230) and is the subject of many commentaries, including that of Thomas Aquinas (Arens 1984, 397–457). The work has been highly influential in the history of Liberal Arts education in the West (for the history of Liberal Arts education, see Conrad, 2014). In this work, Aristotle says that 'words spoken are symbols or signs of affections or impressions of the soul'. Since 'written words are the signs of words spoken', they also symbolise the experience of the soul or psyche. Just as people have different ways of writing, so they also have different speech sounds. He continues, 'But the mental affections themselves, of which these words are primarily signs, are the same.' (Aristotle and Cook 1938, 115) The inner experiences which these things evoke might differ in content or individual experience. Formally, the experiences are the same for all.

In Chapter 2 of *Laws of Form* (2011, 5), Spencer-Brown considers three forms of arrangement ($\overline{\ \ }|\overline{\ \ }|$, $\overline{\overline{\ \ }|\ }|$, $\overline{\ \ }|$) and one absence of form (   ) taken from his primitive equations. Might we be able to map these to the four main sentence types above?

Taking 'Jack and Jill went up the hill' as an example of a declarative sentence, I would symbolize the quality of the relationship between the knower and the known which gives rise to a statement as a mark ($\overline{\ \ }|$). I see a one-to-one or one-to-three correspondence between the knower, that which they know, and the connection between them. The mark distinguishes an inside and an outside and the boundary between them.

The question ('Did Jack and Jill go up the hill?' is an example), however, shows an oscillation between something that is marked and something that is unmarked from the perspective of either the questioner, the listener, or both. It could be one thing; it could be another. Whether or not the questioner knows the answer does not detract from the fact that they have to re-enter the state of the indeterminate relationship in order to ask the question. Until a satisfactory answer is arrived at, the answer is indeterminate. For that reason, I use the mark of recursion to symbolise this particular oscillatory quality of thought which gives rise to the question ($\sqcup$). No other sentence type shares this feature. No other symbol from *Laws of Form* fits it as well.

I see the exclamation ('Jack and Jill went up the hill!') as a *eureka* moment. A *eureka* moment comes in like a flash. You realise the meaning of it. You have to express it and explain it, so the mark expands to two marks ($\overline{\ \ }| \rightharpoonup \overline{\ \ }|\overline{\ \ }|$), something Spencer-Brown introduces as 'confirmation' in his fifth canon: expansion of reference (2011, 8). The order ('Jack and Jill, go up the hill'), however, involves somebody giving a command. To do this, we have to see what the present situation is, what the future could be, and bring both together in the imagination so that when the order is fulfilled, the imagined situation becomes reality and the envisioned future is realized in the present. Once again, Spencer-Brown's fifth canon provides a solution, in the 'condensation' of two marks to one ($\overline{\ \ }|\overline{\ \ }| \rightharpoonup \overline{\ \ }|$). This perfectly depicts the particular quality of thought described which gives rise to a command or an order, as described above.

So far, we have looked at two marks side by side ($\overline{\ \ }|\overline{\ \ }|$) and looked at the expansion and condensation relating to these which Spencer-Brown outlines in his fifth canon which reappears in algebraic form as Consequence 5: Iteration (C5) (2011, 27). We have taken the single mark ($\overline{\ \ }|$), we have added the reentry that is alluded to in Chapter 11 (⊔) and developed by Varela (1975). Now let us consider the mark-over-mark (⏋⏋), which is equivalent to the unmarked state, and the absence of form (   ). Can *Laws of Form* teach us something about sentences?

The mark-over-mark (⏋⏋), I think, maps perfectly to a sentence type that used to be taught. It is called the optative mood and it stands for the wish or prayer. (Daniel 1891, 16; Joseph 2002, 54) The identification, according to Diogenes Laertius, can be traced back to Protagoras (Hicks 1972, 7:66, 9:53). When I say, "God, save the king," I am not ordering God about. I am saying something like 'May God save the king'. It is a wish or prayer. One does not order God about. It is just not done. But the wish or prayer posits something that could be there, maybe should be there, but that is not. The king might not need saving, but should that need arise, then the wish for salvation to occur becomes valid. And we acknowledge that. The mark-over-mark (⏋⏋) seems a perfect fit to denote something that does not have existence but that is wished or prayed for.

What about the absence of form (   )? This is not the same as the mark-over-mark. It is an absence of form. And for me, this describes the very real quality of thought which Dante might have had when faced with his divine vision or the quality of thought I experience when I hear a piece of music performed beyond a level of human mastery, or when a speaker's performance reduces me to tears, or when I am struck speechless by the beauty of an animation or a turn of phrase. While I may have been rendered dumbstruck, yet my quality of thought is vibrant. What better way to depict this very real quality of thought, to acknowledge it and bring it into the discussion than to use Spencer-Brown's symbol for the absence of form (   )?

But these are not the only sentence types that we can recognize. There is a type of question that came in around the time that Spencer-Brown was writing *Laws of Form*: the interrobang ('Jack and Jill did what, now?!'). There is even a special character symbol for the interrobang: (‽). It has a dynamic emotional aspect to it as well as a linear rational aspect to it. There is a symbol from *Laws of Form* which is perfectly suited to this: the double barb (⇌) (Spencer-Brown, 2011, 10). It combines both the oscillatory quality of the question (⊔) and the condensing quality of the exclamation (⏋ ⇀ ⏋⏋). It can also have the quality of a command (⏋⏋ ⇀ ⏋) in it. There is an implied 'Tell me exactly what you mean' in the above interrobang sentence, for example. I use the double barb symbol quite extensively in my book *Story and Structure: A Complete Guide* as explained in the introductory outline (2022, 6).

Spencer-Brown does not elaborate on the double barb in *Laws of Form*, but its use is clearly demonstrated in the work in which it plays an extremely important role.

Seven sentence types arise from taking a different view of sentences. Seeing them not in written form or in terms of verbal content but in terms of different relationships between knower and known, we see sentences as originating in different types of thought or, in Aristotle's terms different 'affections or impressions of the soul' (Aristotle and Cook 1938, 115). The seven types identified to date are outlined and mapped to Spencer-Brown's work in Table 1 below.

Table 1: Sentence types mapped to symbols from George Spencer-Brown's *Laws of Form*

Sentence type	Example	Symbol
Declarative sentence (statement)	Only two can play this game.	┐
Interrogative sentence (question)	Can only two play this game?	⊐
Exclamatory sentence (exclamation)	Only two can play this game!	┐⇀┐┐
Imperative sentence (order or command)	Two, play this game.	┐┐⇀┐
Optative sentence (wish or prayer)	If only two could play this game.	╗
Interrobang sentence (interrobang)	You mean to say that only two can play this game?!	⇌
Awestruck silence ( )	[ ]	

In *Laws of Form*, Spencer-Brown starts out from the first distinction which he sees as a severance. He states that 'by tracing the way we represent such a severance, we can begin to reconstruct, with an accuracy and coverage that appear almost uncanny, the basic forms underlying linguistic, mathematical, physical, and biological science, and can begin to see how the familiar laws of our own experience follow inexorably from the original act of severance.' (2011, xxii). All I have done in the above outline is to take this statement seriously; explore the qualitative differences in the basic states, canons, and consequences outlined in *Laws of Form*; and seek ways in which they might map onto the different qualities of thought between the knower and the known which give rise to the different sentence types we use in the English language. We have come full circle and can now see that the distinctive types of sentences we use are based on basic foundational forms and that, unsurprisingly to those familiar with *Laws of Form*, the process is worthwhile engaging with. As Spencer-Brown puts it, 'Although all forms, and thus all universes, are possible, and any particular form is mutable, it becomes evident that the laws relating such forms are the same in any universe.' (2011, xxiii)

It appears that there are significantly more types of sentences than are taught in school. *Laws of Form* points to the need to reexamine the way we teach language. The optative sentence, the wish or prayer, is an embodied aspect of our thinking. We wish; we pray. Why exclude the optative from the curriculum? The awestruck silence also, is a distinct type of sentence. If we acknowledge it, as Spencer-Brown does, as primary (2011, 85), the act of doing so would help to cultivate people's respect for the universal source of being. I say more about how this applies to sentences put together in paragraphs and narratives in my book *Story and Structure* (2022). I, for one, teach seven sentence types, and I get my students to explore the potential of all seven in creative writing projects. In doing so, they have to confront what it feels like to express a wish or prayer; to evoke a source of awestruck silence. They often come up with some really interesting ways of expressing that blank space.

About the Author

Leon Conrad is affiliated with *The Next Society Institute.* He had the privilege of being Spencer-Brown's last student. He is the founder of *The Academy of Oratory* and *The Traditional Tutor.* He set up *The Unknown Storyteller Project* in 2018 to explore the application of the Calculus of Indications in Spencer-Brown's *Laws of Form* to the analysis of story structure. His book *Story and Structure* has received a number of literary awards, including the IPNE Nonfiction Book of the Year Award 2022 and was shortlisted for The People's Book Prize. For more information, visit his website at https://www.leonconrad.com.

References

[1] Arens, Hans, ed. 1984. *Aristotle's Theory of Language and Its Tradition: Texts from 500 to 1750.* Amsterdam Studies in the Theory and History of Linguistic Science (Series III), vol. 29. Amsterdam & Philadelphia: John Benjamins.

[2] Aristotle. 1938. *On Interpretation.* Translated by Harold P. Cooke. In *The Categories, On Interpretation, Prior Analytics*, translated by Harold P. Cooke and Hugh Tredennick. Cambridge, MA: Harvard University Press.

[3] Australian Curriculum, Assessment and Reporting Authority (ACARA). 2010. *F–10 Curriculum: English (Version 8.4):* Year 1.

[4] Common Core State Standards Initiative. 2010. *Common Core State Standards for English Language Arts & Literacy in History/Social Studies, Science, and Technical Subjects*
https://corestandards.org/english-language-arts-standards/.

[5] Conrad, Leon. 2014. "Integration and the Liberal Arts: A Historical Overview." *On the Horizon* 22 (1): 46–56. https://doi.org/10.1108/OTH-11-2013-0045.

[6] Conrad, Leon. 2020. *Laws of Form Course (Free Online Course).* YouTube playlist. http://youtu.be/playlist?list=PLoK3NtWr5NbqEOdjQrWaq1sDweF7NJ5NB.

[7] Conrad, Leon. 2022. *Story and Structure: A complete guide.* London: Aladdin's Cave Publishing (hbk); Glastonbury: The Squeeze Press (pbk).

[8] Daniel, Rev. Canon Evan. 1891. *The Grammar, History, and Derivation of the English Language.* 2nd rev. ed. London: National Society's Depository. https://archive.org/details/grammarhistoryde00danirich.

[9] Diogenes Laertius. 1925 [1972]. *Lives of Eminent Philosophers.* Translated by Robert Drew Hicks. Cambridge, MA: Harvard University Press. https://www.perseus.tufts.edu/hopper/text?doc=Perseus:text:1999.01.0258.

[10] HM Government. 2013. *Glossary for the Programmes of Study for English (Non-Statutory).* https://www.gov.uk/government/uploads/system/uploads/attachment_data/file/238985/English_Glossary.pdf.

[11] Joseph, Sister Miriam. 2002. *The Trivium: The Liberal Arts of Logic, Grammar, and Rhetoric.* Edited by Margaret McGlinn. Philadelphia, PA: Paul Dry Books.

[12] Spencer-Brown, George. 2011. *Laws of Form.* Leipzig: Bohmeier Verlag. First published 1969 by George Allen & Unwin, London.

[13] Varela, Francisco J., 1975. "A Calculus for Self-reference." *International Journal of General Systems* 2: 5–24.

The Unknown Storyteller Award

As revealed at the 2024 AGM of the Spencer-Brown Society,
we are pleased to announce an annual 'Unknown Storyteller Award' of £100
for a high-quality paper submitted for publication to 'Distinction: Journal of Form'
in which the author applies the story structure methodology outlined in
Leon Conrad's book, *Story and Structure: A complete guide* (2022).

Criteria

1. The award is set up to encourage researchers to apply the methodology outlined in *Story and Structure: A complete guide* and papers must show an application of the methodology. The award seeks to promote well-argued papers which feature novel applications, open up new areas of enquiry, or present critical viewpoints.

2. The potential for applying the methodology is wide and includes current events, media, folklore, narrative, film, poetry, and oratory, and incidents from lived experience.

3. The award is open to all and is not limited by age or academic affiliation.

4. Serving members of the committee would not be eligible to receive an award during their tenure, although this need not stop them from submitting papers using the methodology to the journal or being eligible for to receive an award when they are no longer serving as main or co-opted committee members.

5. Members of the judging panel may submit a paper for consideration. The author will recuse themselves from voting on their own paper.

6. Award will be made on merit only.

7. The paper must use GSB's symbols rather than derivatives such as brackets. A number of resources to include these are available, as featured on the Spencer-Brown Society website at lof50.com/fonts.

8. In the event of there being no entries or if the entries are considered not to be

of high-enough quality, the judges reserve the right to carry over the prize to the following year. The award amount would remain constant.

9. Copyright remains with the author. By submitting the paper for the award, the author grants the Spencer-Brown Society first serial rights for publication of the article, and grants the Spencer-Brown Society the right to publish the article in whole or in part online for marketing and publicity purposes free of charge.

Deadline for submission

Midnight, UK time on 31 December 2025 and annually on 31 December thereafter. The winner will be announced at the Spencer-Brown Society AGM in the following calendar year.

For further information, visit https://lof50.com/award

Diamond Logic
and
How to Count to Two

Nathaniel Hellerstein
paradoctor@aol.com

This essay uses this logic:

```
      T/T
     /   \
  T/F     F/T
     \   /
      F/F
```

to explain this circuit:

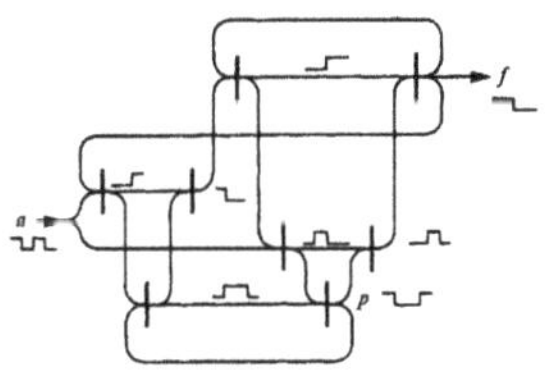

This paper comprises extracts from:

Diamond, A Paradox Logic, 2nd Ed,
N. S. Hellerstein,
Copyright © 2010 by World Scientific Publishing Co. Pte. Ltd.
ISBN-13: 978-981-4287-13-5; ISBN-10: 981-4287-13-X.

and is published here by kind permission.

1 The Liar Paradox

Epimenides the Cretan said that all Cretans lie; did he tell the truth, or not? Let us assume, for the sake of argument, that every Cretan, except possibly Epimenides himself, was in fact a liar; but what then of Epimenides?

He says that he himself lies; but if he is lying, then he is telling the truth; and if he is telling the truth, then he is lying! Which then is it?

The same conundrum arises from the following sentence: "This sentence is false."

That sentence is known as the "Liar Paradox" or "pseudomenon".

The pseudomenon obeys this equation: $L = \text{not } L$.

It's true if false and false if true. Which then is it?

2 Russell's Paradox

Let R be the set of all sets which do not contain themselves:

$$R = \{x | x \notin x\}$$

R is an anti-diagonal set. Is it an element of itself?

In general:

$$x \in R = x \notin x$$

and therefore:

$$R \in R = R \notin R$$

Therefore R is paradoxical. Does R exist?

Bertrand Russell told a story about the barber of a Spanish village. Being the only barber in town, he boasted that he shaves all those—and only those—who do not shave themselves. Does the barber shave himself?

To this legend I add a political postscript. That very village is guarded by the watchmen, whose job is to watch all those, and only those, who do not watch themselves. But who shall watch the watchmen?

Therefore honesty in government is truly imaginary!

3 Santa Sentences

Suppose that a young child were to proclaim: "If I'm not mistaken, then Santa Claus exists."

If one assumes that Boolean logic applies to this sentence, then its mere existence would imply the existence of Santa Claus!

Why? Well, let the child's statement be symbolized by 'SS', and the statement "Santa exists" be symbolized by 'S'. Then we have the equation:

$$SS = \text{ if } SS \text{ then } S = SS \implies S = \sim SS \vee S.$$

Then we have this line of argument:

S is either true or false. If it's true, then so is SS:

$$SS = (\sim SS) \vee \text{true} = \text{true}.$$

No problem. But if S is false, then SS becomes a liar paradox:

$$SS = (\sim SS) \vee \text{false} = \sim SS.$$

If S is false, then SS is non-boolean. Therefore: If SS is boolean, then S is true.

Note that both arguments work equally well to prove any other statement besides S to be true; one need merely display the appropriate "Santa sentence." Thus, for instance, if some skeptic were to declare: "If I'm not mistaken, then Santa Claus does not exist"—then an identical argument proves that Santa Claus does *not* exist!

Given two opposite Santa sentences:

$$SS_1 = (SS_1 \implies S);\ SS_2 = (SS_2 \implies \sim S)$$

then at least one of them must be paradoxical.

One can tell barber-like stories about Santa statements. For instance, in another Spanish village, the barber takes weekends off; so he shaves all those, and only those, who shave themselves only on the weekend:

> B shaves M = If M shaves M, then it's the weekend.
> One fine day someone asked: does the barber shave himself?
> B shaves B = If B shaves B, then it's the weekend.
> Has it been weekends there ever since?

4 Cantor's Paradox

Cantor's proof of the "uncountability" of the continuum relies on a paradoxical anti-diagonal. For assume that we had a countable list of the real numbers between 0 and 1:

$$R1 = 0\ .\ D_{11}, D_{12}, D_{13}, D_{14} \dots$$
$$R2 = 0\ .\ D_{21}, D_{22}, D_{23}, D_{24} \dots$$
$$R3 = 0\ .\ D_{31}, D_{32}, D_{33}, D_{34} \dots$$

where D_{MN} is the Mth binary digit of the Nth number.

Then we define Cantor's anti-diagonal number:

$$C = 0\ .\ \sim D_{11}, \sim D_{22}, \sim D_{33}, \sim D_{44} \dots$$

If $C = R_N$ for any N, Then $D_{NX} = \sim D_{XX}$;

Therefore $D_{NN} = \sim D_{NN}$; the pivot bit buzzes.

From this paradox, Cantor deduced that the continuum has too many points to be counted—a higher infinity. A single buzzing bit implies infinities beyond infinity. Was more ever made from less?

I say, why seek "transfinite cardinals," whatever those are? Why not ask for Santa Claus? In this spirit, I introduce the Santa-diagonal number:

$$S = 0\ .\ (D_{11} \implies \text{Santa}), (D_{22} \implies \text{Santa}), (D_{33} \implies \text{Santa}) \ldots$$

If $S = R_M$ for any M, then $D_{MX} = (D_{XX} \implies \text{Santa exists})$;

Therefore $D_{MM} = (D_{MM} \implies \text{Santa exists})$.

If the pivot bit is boolean, then Santa Claus exists!

5 Gödel's Paradox

Gödelian metamathematics requires coding statements about mathematical proof as statements about numbers. Gödel used prime factorization coding; given an integer's prime factorization:

$$N = (2^\wedge e_1)(3^\wedge e_2)(5^\wedge e_3)(7^\wedge e_4) \ldots (p_n^\wedge e_n)$$

then N encodes the sequence $e_1, e_2, e_3, e_4, \ldots, e_n$. That can encode any sequence of symbols whose formal properties we can analyze. With Gödel coding we can discuss sentence form, by discussing numbers.

His coding is inefficient. The 21st century way to encode symbol sequences with integers is by binary expansion:

$$N = 110010101100010101 \ldots 111010100001011110$$

If S is a statement written in some formal language, then let 'S' denote the number resulting from writing down S, translating its symbols into ASCII code, and read in binary.

Manipulating binary sequences is routine in assembler, and so we can write computer-code statements about the formal structure of sentences. For instance, we can define predicates for provability and possibility:

$$\begin{aligned} \text{prv}(`S\text{'}) &= S \text{ is provable} \\ &= \text{there exists a number encoding a proof of } S \\ \text{poss}(`S\text{'}) &= S \text{ is possible} \\ &= \text{not prv(`not } S\text{')} \end{aligned}$$

Now consider this self-quoting statement:

" 'is unprovable when preceded by its own quotation' is unprovable when preceded by its own quotation."

or in other words:

"This sentence is unprovable."

G = not prv ('G')

That is "Gödel's Paradox." It says "doubt this sentence." If proof = truth, then Gödel's paradox is a liar paradox. Therefore in Boolean logic there is a disconnect between proof and truth. Boolean logistic fails.

Either G is true, and prv('G') is false, G is true but unprovable, and the proof system is incomplete; or G is false, and prv('G') is true, and the proof system proves falsehoods. Worse, it proves contradictions, because if G = false and prv(G) = true then

$$\begin{aligned} \text{True} &= \sim G &&= \text{prv(`}G\text{')} = \text{prv(prv(`}G\text{'))} \text{ ; but also} \\ \text{True} &= \sim G &&= \text{prv(`}G\text{')} = \text{prv(}\sim \text{prv(`}G\text{')) by definition of } G\text{; so} \\ \text{True} &= \sim G &&= \text{prv(}\sim \text{prv(`}G\text{') and prv(`}G\text{'))} \end{aligned}$$

If $G =$ false, then contradiction is provable, and the proof system fails. So either the system is incomplete or it's inconsistent. That is Gödel's Incompleteness Theorem.

6 Diamond Logic

Start with the Liar Paradox: $L = \sim L$. If $L =$ true then $L =$ false, and if $L =$ false then $L =$ true. This suggests a period-2 wave: $TFTFTFTF\ldots$. Now consider all period-2 waves:

$$\begin{aligned} T/T &= T \\ F/F &= F; \text{ these are the ``Boolean'' values.} \\ T/F &= I \\ F/T &= J; \text{ these are the ``paradox'' values.} \end{aligned}$$

They form a diamond-shaped lattice:

T

I J

F

This is "diamond logic." In diamond logic, the positive operators $\wedge$ ("and") and $\vee$ ("or") operate side-wise:

$$\begin{aligned} (a/b) \wedge (c/d) &= (a \wedge c)/(b \wedge d) \\ (a/b) \vee (c/d) &= (a \vee c)/(b \vee d) \end{aligned}$$

Negation flips sides:

$$\sim (a/b) = (\sim b)/(\sim a)$$

This creates fixed points for negation:

$$\sim (t/f) = (\sim f)/(\sim t) = t/f$$
$$\sim (f/t) = (\sim t)/(\sim f) = f/t$$

Thus paradox is possible in diamond logic.

Call a function "harmonic" if it can be defined from $\wedge, \vee, \sim$, and the four values t, i, j, f. They include:

$$\begin{array}{llll}
a \text{ nor } b & = \sim (a \vee b) & = & \sim a \wedge \sim b \\
a \text{ nand } b & = \sim (a \wedge b) & = & \sim a \vee \sim b \\
a \text{ but } b & = a/b = (a \wedge i) \vee (b \wedge j) & = & (a \vee j) \wedge (b \vee i) \\
M(a,b,c) & = (a \wedge b) \vee (b \wedge c) \vee (c \wedge a) & = & (a \wedge b) \vee (b \wedge c) \vee (c \wedge a) \\
D(x) & = x \vee \sim x & & \\
d(x) & = x \wedge \sim x & &
\end{array}$$

Here are diamond's truth tables for 'not,' 'and,' 'or,' and 'but':

x:	**$\sim$ x:**	**$\wedge$ y:** t f i j	**$\vee$ y:** t f i j	**but y:** t f i j
t	f	t f i j	t t t t	t i i t
f	t	f f f f	t f i j	j f f j
i	i	i f i f	t i i t	t i i t
j	j	j f f j	t j t j	j f f j

7 Diamond Circuits

Logic circuits can operate in diamond logic, via the "dual-rail" system. Replace each wire by a pair of wires, and each gate by a pair of gates. AND and OR gate-pairs operate separately on the sides; NOT, NOR and NAND gate-pairs swap sides after operating.

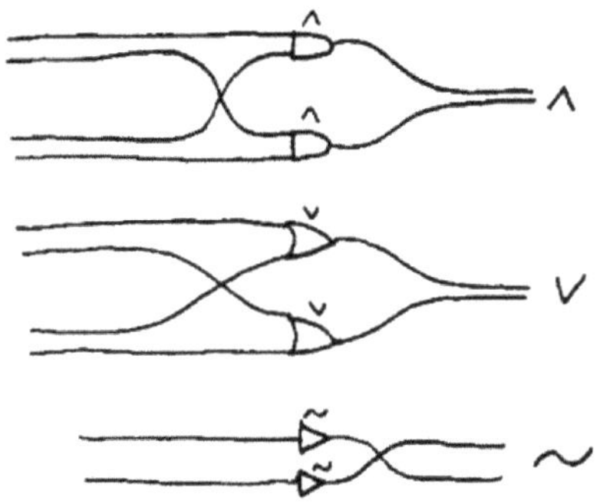

8 Diamond Laws

Diamond logic obeys the **De Morgan Laws:**

Commutativity:	$A \vee B = B \vee A$; $A \wedge B = B \wedge A$
Associativity:	$(A \wedge B) \wedge C = A \wedge (B \wedge C)$
	$(A \vee B) \vee C = A \vee (B \vee C)$
Distributivity:	$A \wedge (B \vee C) = (A \wedge B) \vee (A \wedge C)$
	$A \vee (B \wedge C) = (A \vee B) \wedge (A \vee C)$
Identities:	$A \wedge t = A$; $A \vee f = A$
Attractors:	$A \wedge f = f$; $A \vee t = t$
Recall:	$A \wedge A = A$; $A \vee A = A$
Absorption:	$A \wedge (A \vee B) = A$; $A \vee (A \wedge B) = A$
Double Negation:	$\sim (\sim A) = A$
De Morgan:	$\sim (A \wedge B) = (\sim A) \vee (\sim B)$
	$\sim (A \vee B) = (\sim A) \wedge (\sim B)$

These are the Boolean laws, minus the Law of the Excluded Middle.

In addition to these laws, diamond logic has this 'anti-boolean' law:

Complementarity: $Di \wedge Dj = f \;\; ; \;\; di \vee dj = t$

That is: $(i \vee \sim i) \wedge (j \vee \sim j) = f$

$(i \wedge \sim i) \vee (j \wedge \sim j) = t$

Complementarity defies the Law of the Excluded Middle, but not directly enough to collapse logic to a single value. Instead, it shares contradiction between two paradox values.

These diamond logic axioms suffice to prove this:

Theorem: *Normal Form Theorem*

if $F(x)$ is defined in diamond logic, then for all x:

$$\begin{aligned} F(x) &= (F(T) \wedge x) \vee (F(F) \wedge \sim x) \vee M(F(i), dx, F(j)) \\ &= (F(T) \vee \sim x) \wedge (F(F) \vee x) \wedge M(F(i), Dx, F(j)) \end{aligned}$$

This in turn implies:

Theorem: *Complete Deduction*

The De Morgan laws, plus Complementarity, prove all identity equations in diamond logic.

Define *phase order* $<$ on diamond by this diagram:

```
          T
      <       <
  I       <       J
      <       <
          F
```

Phase order is **preserved by harmonic functions**:

If $F(x_1, \dots x_N)$ is defined by $\wedge$, $\vee$, $\sim$, variables, and constants

and $x_1 < y_1,\ x_2 < y_2, \dots,\ x_N < y_N$

then $F(x_1, \dots x_N) < F(y_1, \dots y_N)$

This implies, by iteration from $(i, i, i, \dots i)$ or from $(j, j, j, \dots, j)$:

Theorem: *Complete Self-Reference*

If $F(x)$ is a n-component harmonic function on n variables:
$F(x) = (F_1(x_1, \dots, x_n), \dots, F_n(x_1, \dots, x_n))$
then F has a fixed point, that is, a vector $s = (s_1, \dots, s_n)$ such that
$F(s) = s$

9 Paradox Resolved

Liar Paradox: $L = \sim L$
This is solved by $L = i$ or $L = j$.

Santa Sentence: $SS = (SS$ implies false$) = \sim SS$
This is solved by $SS = i$ or $SS = j$.

Russell's Paradox: $(R \in R) = (R \notin R)$
This is solved by $(R \in R) = i$ or $(R \in R) = j$.

Gödel's Paradox: if proof = truth then $G = \sim G$
This is solved by $G = i$ or $G = j$.

In boolean logic, these have no solutions; in diamond logic, they have two solutions. The choice of i or j conveys *phase information*.

10 Systems, Ganglions, and Flexware

Consider this system, the Triplet:

$$X = \sim (Y \vee Z)$$
$$Y = \sim (Z \vee X)$$
$$Z = \sim (X \vee Y)$$

This has the circuit:

It has five stable states, arranged in an M_3 lattice:

		TFF		
	/		\	
III	–	FTF	–	JJJ
	\		/	
		FFT		

Its defining equations can be written with multiplex plus negations:

$$X = \sim ((X \wedge f) \vee (Y \wedge t) \vee (Z \wedge t))$$
$$Y = \sim ((X \wedge t) \vee (Y \wedge f) \vee (Z \wedge t))$$
$$Z = \sim ((X \wedge t) \vee (Y \wedge t) \vee (Z \wedge f))$$

Systems in general can be written with multiplex plus negations:

$$\begin{aligned} X_1 &= \sim ((X_1 \wedge C_{11}) \vee \ldots \vee (X_N \wedge C_{1N})) \\ \ldots \\ X_N &= \sim ((X_1 \wedge C_{N1}) \vee \ldots \vee (X_N \wedge C_{NN})) \end{aligned}$$

This defines a circuit with this circuit matrix:

$$\begin{matrix} C_{11}, \ldots, & C_{1N} \\ \ldots \\ C_{N1}, \ldots, & C_{NN} \end{matrix}$$

The Triplet has this circuit matrix:

F, T, T
T, F, T
T, T, F

You can combine a multiplexer, written in dual-rail, with a row of dual-rail negations, to construct this circuit, the Ganglion:

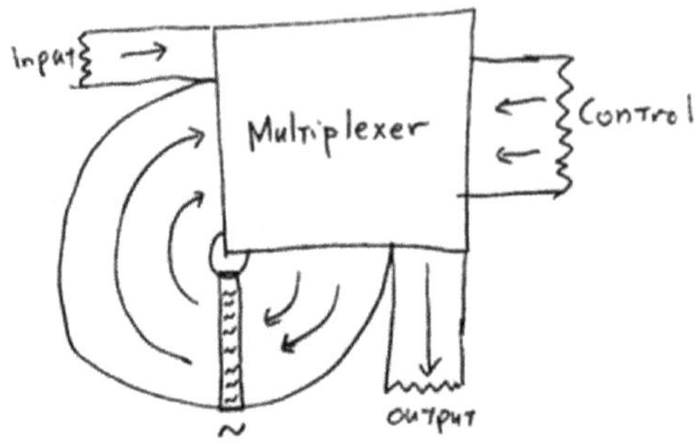

The ganglion can emulate any diamond-logic circuit. That circuit can be quickly changed. This is "flexware"—a level of computation between hardware and software.

Flexware lends itself to optimization by genetic algorithms. Circuits can be mutated, combined, and hybridized by the J gate (see below).

The following pages display these flexwares: Rotors, Pumps, and Tapes.

11 Modulators to Rotors

Here is the "**First Brownian Modulator**":

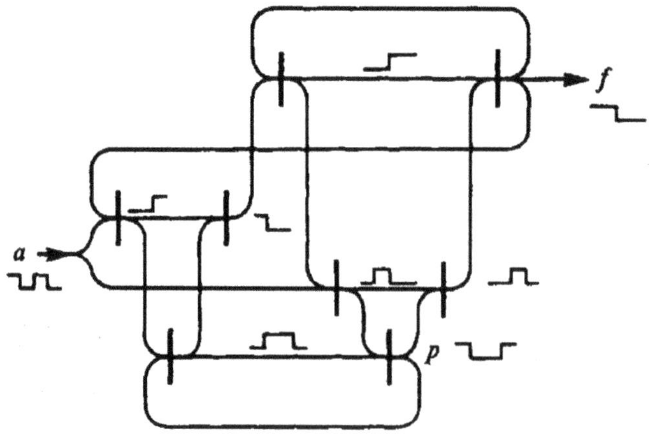

Its function is to count to two. Two oscillations of the A input yield one oscillation of the F output.

It equals this system:

$$
\begin{aligned}
a &= \sim(z \vee d) \\
b &= \sim(a \vee g) \\
c &= \sim(d \vee b) = z/2 \\
d &= \sim(c \vee f) \\
e &= \sim(c \vee z) \\
f &= \sim(e \vee h) \\
g &= \sim(a \vee h) \\
h &= \sim(e \vee g)
\end{aligned}
$$

In this circuit diagram, the triangles denote NOR gates:

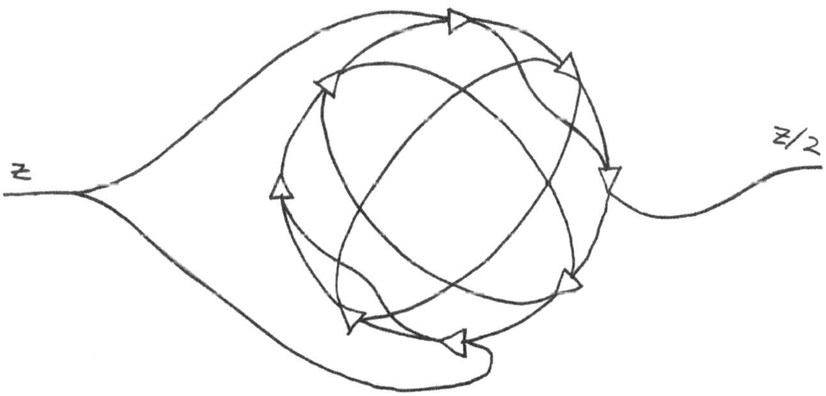

$z/2$ toggles once when z toggles twice.

From the juxtaposition operator /, define the *intermix* function:

$$J(a,b) = (a/b, b/a)$$

Note: $(J(J(a,b)) = (a,b)$
$\sim J(a,b) = J(\sim b, \sim a)$

With termwise ∨, ∧, and nor:

$$J(a,b) \vee J(A,B) = J(a \vee A, b \vee B)$$
$$J(a,b) \wedge J(A,B) = J(a \wedge A, b \wedge B)$$
$$J(a,b) \text{ nor } J(A,B) = J(b \text{ nor } B, a \text{ nor } A)$$

Let $(A,E) = J(a,e)$; so that $A = a/e$, and $E = e/a$.
Also let $(B,F) = J(b,f); (C,D) = J(c,d); (G,H) = J(g,h)$
This is "diffracting the modulator against itself."

Then:

$$\begin{aligned} A &= \sim(C \vee z) \\ E &= \sim(D \vee z) \\ B &= \sim(E \vee H) \\ F &= \sim(A \vee G) \\ C &= \sim(C \vee F) \\ D &= \sim(D \vee B) \\ G &= \sim(G \vee E) \\ H &= \sim(H \vee A) \\ z/2 &= C/D \end{aligned}$$

Write this in a cycle to get a "Brownian Rotor":

$$\begin{aligned} A &= \sim(C \vee z) \\ H &= \sim(A \vee H) \\ B &= \sim(H \vee E) \\ D &= \sim(B \vee D) \\ E &= \sim(D \vee z) \\ G &= \sim(E \vee G) \\ F &= \sim(G \vee A) \\ C &= \sim(F \vee C) \\ z/2 &= C/D \end{aligned}$$

In this diamond circuit diagram, J denotes an intermix gate. The table shows four stable states of the system, which are successive in a period-4 cycle, when the input z toggles between T and F four times.

z	A	H	B	D	E	G	F	C	C/D
T	F	i	i	i	F	j	j	j	F
F	j	j	F	i	i	i	F	j	F
T	F	j	j	j	F	i	i	i	T
F	i	i	F	j	j	j	F	i	T

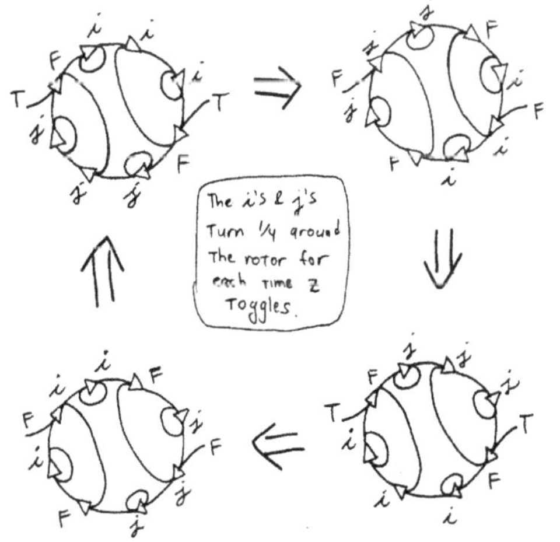

Now consider "Kauffman's Modulator" (Kauffman, 1994):

$$\begin{aligned} a &= \sim(b \vee d \vee z) \\ b &= \sim(a \vee e) \\ c &= \sim(d \vee f) \\ d &= \sim(a \vee c \vee z) \\ e &= \sim(a \vee f) = z/2 \\ f &= \sim(d \vee e) \end{aligned}$$

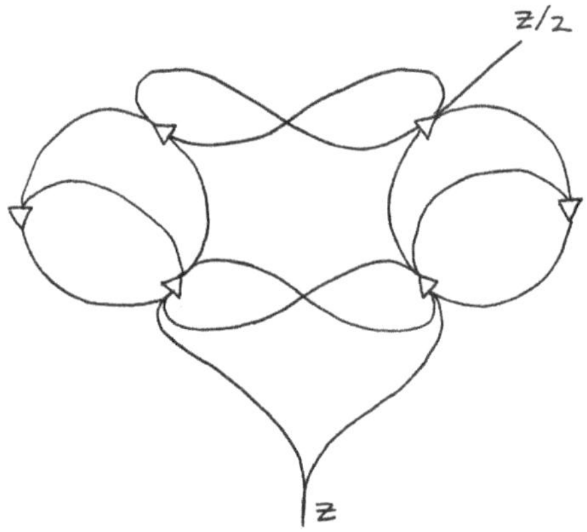

Let $(A, D) = J(a, d)$; so that $A = a/d$, and $D = d/a$.
Also let $(B, C) = J(b, c)$; $(E, F) = J(e, f)$
This too is diffracting the modulator against itself.

Then:

$$\begin{aligned} A &= \sim(A \vee C \vee z) \\ D &= \sim(D \vee B \vee z) \\ B &= \sim(D \vee F) \\ C &= \sim(A \vee E) \\ E &= \sim(D \vee E) \\ F &= \sim(A \vee F) \\ z/2 &= E/F \end{aligned}$$

Write this in a cycle to get a “Kauffman Rotor" (ibid.):

$$
\begin{aligned}
A &= \sim(C \vee A \vee z) \\
F &= \sim(A \vee F) \\
B &= \sim(F \vee D) \\
D &= \sim(B \vee D \vee z) \\
E &= \sim(D \vee E) \\
C &= \sim(E \vee A) \\
z/2 &= E/F
\end{aligned}
$$

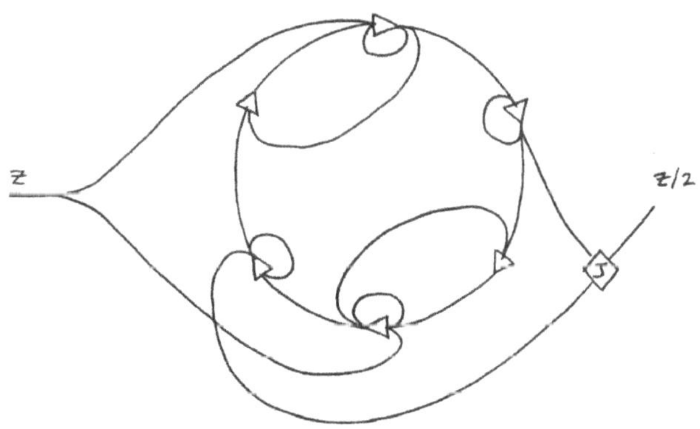

z	A	F	B	D	E	C	E/F
F	i	i	F	j	j	F	F
T	F	i	i	F	j	j	F
F	j	j	F	i	i	F	T
T	F	j	j	F	i	i	T

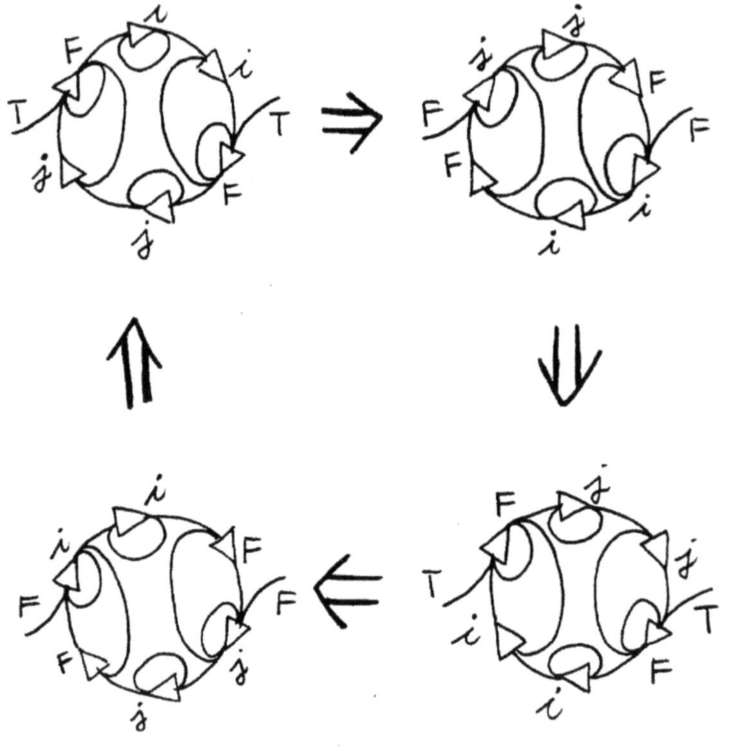

12 Pumps and Tapes

Note that these modulators, when diffracted, become rotors! The i's and j's chase each other around a circle.

All these circuits require correct initialization; they need both i's and j's to have "traction." I therefore suggest adding special initialization leads into these circuits.

Note how the states are self-sealing; even when z is off, the i's and j's set up roadblocks against each other. This is because $\sim (i \vee j) = F$. These circuits demonstrate the purpose of the Second Paradox: to encode phase information.

If these diamond circuits are dual-rail, then each gate is double. Thus, Brownian rotors require 2*8 = 16 gates, and Kauffman's rotor requires 2*6 = 12 gates. The circuit takes up to 4 steps per input flip. Ergo circuit cannot halve frequency unless it is at least 4 times faster than input.

If we lay out the Brownian or Kauffman rotors out in a line, then we get "pumps", or "one-way tapes".

Brownian pump:

$$
\begin{aligned}
&\dots \\
A_n &= \sim(C_n \vee D_{n-1}) \\
B_n &= \sim(A_n \vee B_n) \\
C_n &= \sim(B_n \vee Z) \\
D_n &= \sim(C_n \vee D_n) \\
&\dots
\end{aligned}
$$

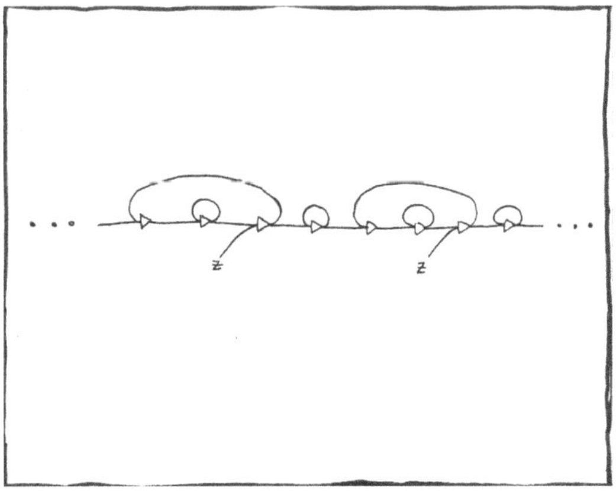

Kauffman pump (ibid.):

$$\begin{aligned} &\ldots \\ A_n &= \sim (A_n \vee C_{n-1}) \\ B_n &= \sim (A_n \vee C_n) \\ C_n &= \sim (B_n \vee C_n \vee Z) \\ &\ldots \end{aligned}$$

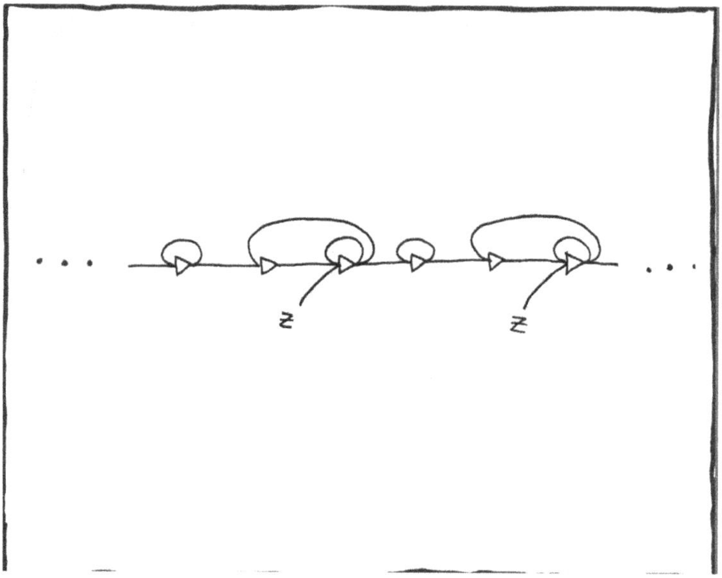

In these circuits, the i's and j's travel one block down the one-way tape when Z toggles twice.

Adding a d variable, for direction, makes "two-way tapes":

Brownian tape:

$$
\begin{aligned}
&\dots \\
A_n &= \sim(A_n \vee E_{n-1} \vee D_{n+1})) \\
B_n &= \sim(A_n \vee D_n \vee E_n) \\
C_n &= \sim(B_n \vee C_n) \\
D_n &= \sim(C_n \vee \sim d \vee Z) \\
E_n &= \sim(C_n \vee d \vee Z) \\
&\dots
\end{aligned}
$$

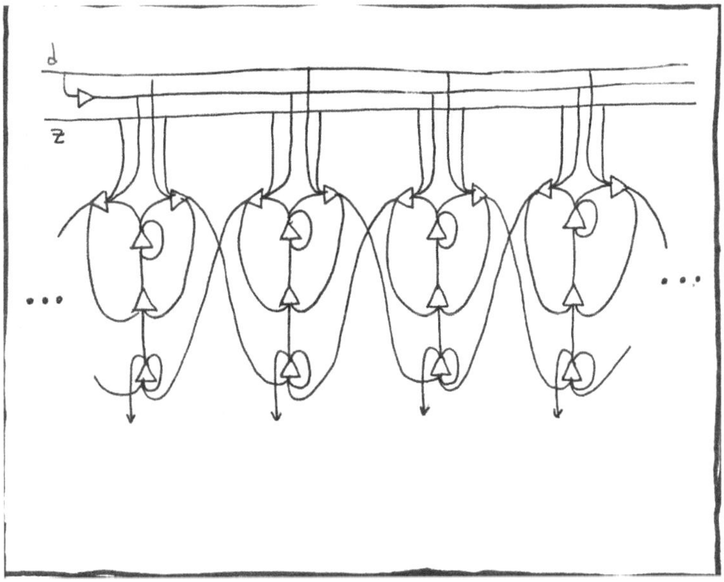

Kauffman tape:

$$
\begin{aligned}
&\dots \\
A_n &= \sim(A_n \vee D_{n-1} \vee C_{n+1}) \\
B_n &= \sim(A_n \vee D_n \vee C_n) \\
C_n &= \sim(B_n \vee C_n \vee \sim d \vee Z) \\
D_n &= \sim(D_n \vee B_n \vee d \vee z) \\
&\dots
\end{aligned}
$$

In these circuits, the i's and j's travel one block along the tape, in the direction indicated by d, when Z toggles twice.

Two complementary tapes, juxtaposed, yield a binary tape; connect them to a finite-state machine which feeds back; that makes a Turing machine.

References

[1] Hellerstein, Nathaniel S. 1984. *Diamond: A Four-Valued Approach to the Problem of Paradox.* PhD dissertation, University of California, Berkeley.

[2] Hellerstein, Nathaniel S. 1997a. *Diamond: A Paradox Logic.* Singapore: World Scientific.

[3] Hellerstein, Nathaniel S. 1997b. *Delta: A Paradox Logic.* Singapore: World Scientific.

[4] Hellerstein, Nathaniel S. 2010. *Diamond: A Paradox Logic.* 2nd ed. Singapore: World Scientific.

[5] Kauffman, Louis H. 1994. "Knot Automata." In *Proceedings of the 24th International Symposium on Multiple-Valued Logic (ISMVL 1994)*, 328–333. IEEE Computer Society. https://doi.org/10.1109/ISMVL.1994.302182.

[6] Kauffman, Louis H., and Francisco J. Varela. 1980. "Form Dynamics." *Journal of Social and Biological Structures* 3 (2): 171–206. https://doi.org/10.1016/0140-1750(80)90008-1

[7] Spencer-Brown, George. 1979. *Laws of Form*. New York: The Julian Press.

Flourishing of the Form—Co-arising of Possibility Fields

Walter Tydecks
Independent scholar
walter@tydecks.info

Abstract

Can the idea of an interdependent co-arising of form be seen in the *Laws of Form*? Spencer-Brown's "Construction: Draw a Distinction" (Spencer-Brown 2011, 3) leaves open who is speaking here, who is being addressed and what kind of conversation it is. It contains four moments: the medium (the formable, distinguishable, *void*), in which the form is inscribed; the drawer (the distinguisher), who is spoken to with this request; the process of forming (drawing, distinguishing); the form (the sign, the difference), which is produced as a result. In the sense of interdependent co-arising, the four moments are interdependent and can only arise together (cooperatively). Each of the moments is both active and passive. Western thought, on the other hand, sees the activity exclusively in the subject (the drawer, the distinguisher). For Western thinking, therefore, the idea that the medium is also active and not merely a passive material into which something is inscribed is especially unfamiliar. Equally uncommon is the idea that the resulting form has an activity of its own, with which it can both act back on the drawer and affect others who are attracted or repelled by that form. Even the process of drawing (distinguishing) has its own process form that can develop its own dynamics (second-order forms).

I want to thank Leon Conrad, Fred Cummins and Claire Hill for their assistance with the translation into English. They not only suggested the appropriate English words, but also contributed to greater clarity of content.

1 Question, thesis and objective of the article

Can a flourishing of form emerge from the rather statically conceived *laws* of form, whereby the energy of the form shows itself? Can the observer leave his distance from the forms he observes and enter into an open interaction with them, mediated through the medium? The presentation is divided into five sections.

Mathematical category theory has developed a framework for mapping the properties of static objects and the properties of their motions onto each other. This will be exemplified by mechanical spaces of motion (tangent bundles), quantum mechanical symmetry changes, and Spencer-Brown's relation of first- and second-degree equations.

The idea of interdependent co-arising goes back to Aristotle and Nagarjuna (the Indian philosopher of Mahayana Buddhism): A movement does not arise unilaterally by the effect of an active subject on a passive object, but from the interaction of the movable, the movable-capable and the resulting process of movement.

With the theory of evolution, a paradigm shift from a strict, deterministic movement to an open, selecting movement has taken place, in which possibility fields arise for both the movable and the moving-capable, from which the process of movement dynamically emerges. This is discussed by Ellerman (2007) in the language of category theory under the heading of 'adjoint functor pairs.'

These results are applied to Spencer-Brown's *Laws of Form*. The three movements of forming, drawing and observing mutually condition each other.

From this the result is given: If, in a rigid system, the dynamizable and the dynamizing-capable find each other, the flourishing of the rigid system into liveliness arises from their interaction.

The paper finishes with an example from the formation history of the arrow, which has become the most important operator in category theory, taking the place of statically conceived numbers.

Note: This article is based on a lecture at the *Laws of Form Conference 2022* (LoF22) on August 5, 2022 in Liverpool, which was repeated on September 21, 2022 in a Hegel reading group at the University of Tübingen. Two paragraphs and two chapters have been added in response to the discussions and suggestions.

2 Category Theory of Mechanical Systems

The *flourishing* of a form is a movement that comes from within, shows itself outwardly and causes a feeling of joy for every observer. Is there a formal figuration for this? For me, no work on logic shows such exuberant joy as *Laws of Form*. Heinz von Foerster is said to have laughed out loud with enthusiasm again and again when he first read this book. What is the relationship between logic and joy?

First of all, the outdated idea of numbers, according to which they are nothing but unchanging quantities without inner movement and development, must be abandoned. Mathematics and logic had started with laws which were valid independent of space and time. The turn came with the study of mechanical movements. While earlier, the properties of *given* things were mainly investigated like their color, smell, weight etc. and, of numbers, static properties such as their divisibility, arrangement, and their equivalence classes—such as even and odd numbers—were sought, with the experiments of mechanics, *variable quantities* arose, the differentials, which have their own calculus. For a long time, the way in which modes of calculation with ordinary numbers, and higher analysis with differentials could be combined remained an open question. The answer arose at the beginning of the 20th century, when it was possible to describe the interaction of both calculi using graphical methods. This is best illustrated by the example of non-linear curves.

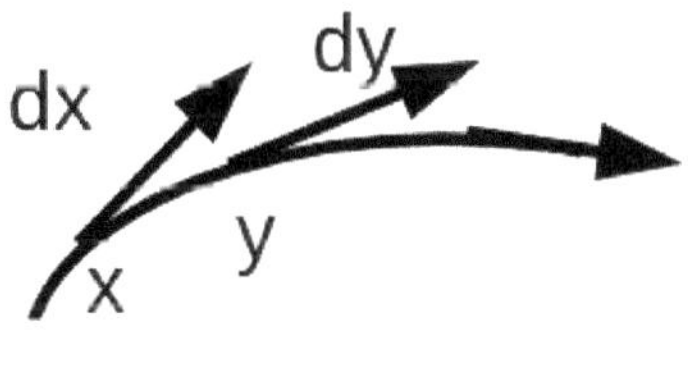

Figure 1: Tangents along a curve

On one side there is a non-linear curve, the points of which are marked with x, y; on the other side, there is the sequence of the respective straight-line tangents dx, dy, which are applied to these points. In mechanics, dx, dy correspond to the respective instantaneous velocities along a path of motion at the points x, y. The differential calculus agrees with the ordinary calculus if the independent operations in both calculi come together again in the result and agree.

The tangents along a continuous, curved line constitute an inner connection. If a transition is made from one tangent dx to an adjacent tangent dy, this should correspond to a transition between the two continuously connected points x and y at which the tangents are applied (Fig. 2). This example employs the method of category theory to represent this in two dimensions:

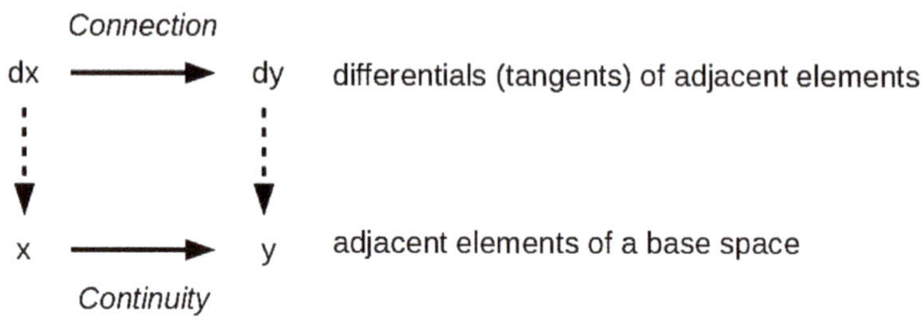

Figure 2: Tangential bundle, commutative diagram

Mathematically, we speak of a commutative diagram in which two paths can be interchanged (commutated). This idea was extended during the 20th century. Not only tangents along a curve are considered, but also cotangents, local coordinate systems and symmetries (which are called fibers in general).

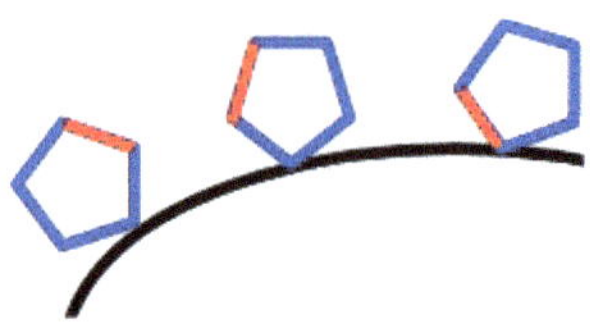

Figure 3: Fibers along a curve

In this example (Fig. 3), a regular pentagon is chosen as a symmetry that moves along a curve. The pentagon can not only change its position from x to y in the course of the movement, but can also rotate with its own movement (*Eigenbewegung*) and change the symmetry properties of its outer sides. According to this idea, mathematically much more complex formulas are used to describe the properties of different forms of motion, from mechanical motion to particle physics and molecular biology. For each form of motion, a certain symmetry is valid. As with the example of the tangents, it should be the case that the transition from one symmetry into an adjacent symmetry corresponds to the transition of adjacent points of a continuous function (Fig. 4).

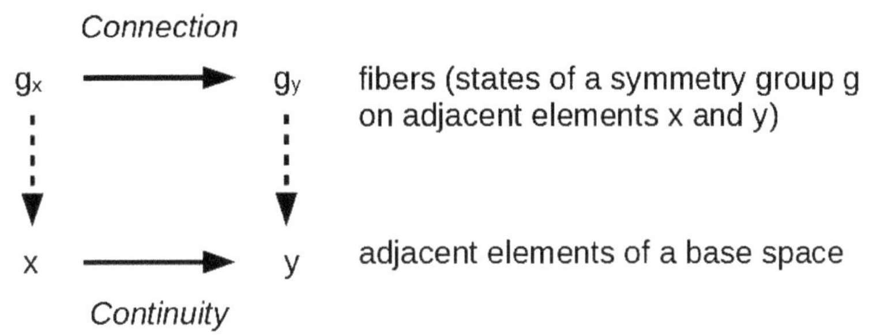

Figure 4: Fiber bundle, commutative diagram

Is it possible to transfer this concept to Spencer-Brown's *Laws of Form*? Without going into this in more details here, suffice it to say that the first and second degree forms introduced by Spencer-Brown and his successors can be brought into a similar relationship:

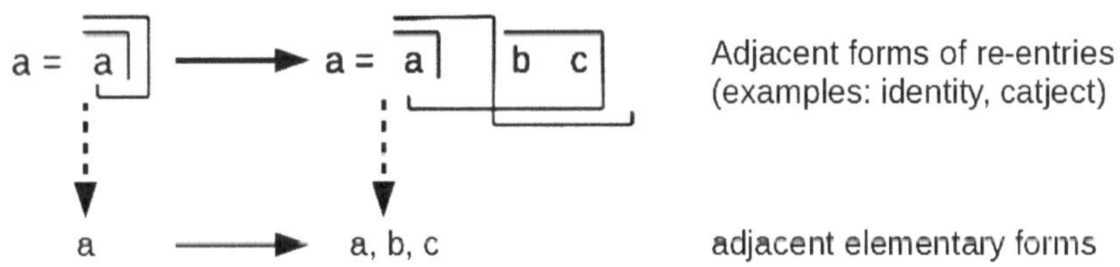

Figure 5: Forms and re-entries, commutative diagram

In the lower line, the letters *a*, *b* and *c* represent forms of the first degree (Fig. 5). These forms stand side by side, as is familiar from ordinary numbers. Spencer-Brown has demonstrated, with second-degree equations, how first-degree forms can give rise to infinite sequences, which, comparable to ordinary differential calculus, lead via a boundary transition into higher forms (second-degree forms) that contain a re-entry. The infinite movement of the return of the re-entry corresponds to the differentials (in Newton's expression, the fluxions). The relationships between the re-entries are comparable to the differential calculus. We consider some initial examples of how re-entries are to be understood. For an elementary form a, a re-entry can be formed that describes the going out of the form a into its other and the return from the other to itself. For three elementary forms *a*, *b* and *c*, a catject can be formed across the board, as recently introduced by Dirk Baecker (Baecker 2001, 14, 57). Category theory can be used to ask whether there is a connection between second-degree equations that corresponds to the connection between fibers.

3 Co-Arising (Aristotle, Nagarjuna)

Using the methods of category theory, ideas of the Indian philosopher Nagarjuna (ca. 2nd century) and Aristotelianism can be formalized and combined. From Nagarjuna, we take the insight that, contrary to Western thinking, no movement is a unilateral effect of a subject on an object (of an agent on his given item), but is composed of three moments that mutually condition each other. The classic example of this is the movement of walking. It can be formalized using the methods of category theory (Fig. 6):

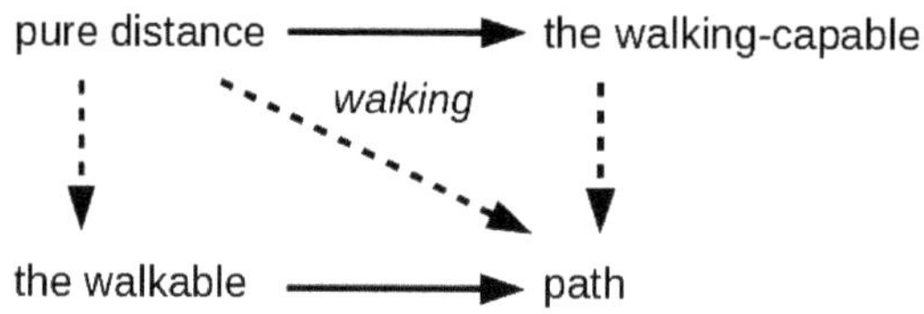

Figure 6: From a distance to the path

There can be no walking if there is no one who can walk and if there is no path that can be walked. Conversely, there can be no walker if there is no walking as a form of movement and if there is no path to walk. And finally, there can be no distance to be walked if there is no one to walk it and no form of movement of walking. This paradox can be generalized from the example of walking and transferred to counting, speaking, perceiving, distinguishing, forming, inscribing something in a ground, and so on.

Viewed as co-arising, we no longer simply describe the *results obtained* (as, in this example, the path), but the *process of becoming*. Co-arising follows its own process logic, which differs from the static logic of the given things. The process is an in-between (in German *mittendrin*, in Greek *metaxy*), between a beginning and an end, where it is still open whether the development will succeed and how it will look like in the result. The different moments of emergence are in flux and change each other. This becomes clearest with the number. Before there can be a number, there must be something that *can* be counted and something that *can* count. Both are initially only possibilities that must come together, and whether counting will succeed is initially open.

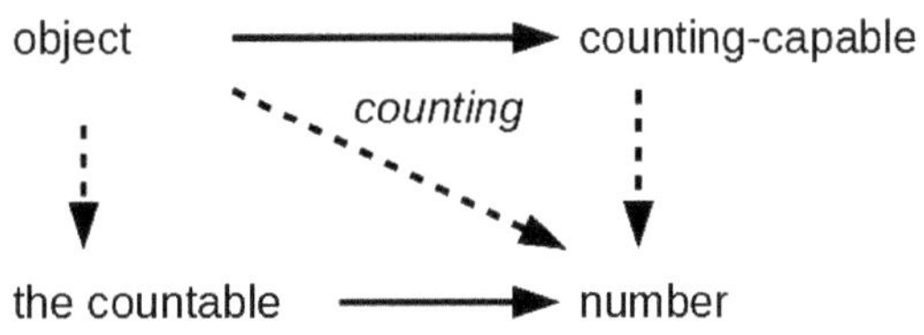

Figure 7: From a counted object to a number

The countable and the counting-capable (and, by analogy, the formable and the forming-capable, the distinguishable and the distinguishing-capable etc.) are not externally opposed, but are in an interaction (Fig. 7). The formable is, so to speak, awakened (aroused) in its formability by the forming-capable, and, vice versa, the forming-capable by the formable. One who can form, learns this ability only if he encounters the formable. The formable triggers in him the ability of being able to form. This must already be present, but is activated only by the encounter with something formable.

I take the internal distinctions of number from Jeck (1994), which he compiled from Aristotle's commentaries from Boethus of Sidon (in the 1st century AD) to Simplicius (480-560): We can distinguish the countable (everything that can be counted) (*arithmeton*), the countable itself (the specific individual object that is countable) (*arithmeton auto*), the counted (that which is counted) (*arithmoumenon*), the process of counting (*arithmeisthai*), the being that will count (*arithmesontos*), and the counting being (*arithmoun*) (Jeck 1994, 42).

Addendum 1: Margaretha Hendrickx asked: "When speaking of the field of possibility of the capable, who is meant by the capable?" In my understanding, first and foremost, of course, the specific human being who is capable of doing something. But it can also be a machine (e.g. the Turing machine, in which Wittgenstein saw an independent *form of life [Lebensform]* of the speaking-capable), the unconscious within us, the collective unconscious, a demon such as Descartes', the divine, etc.

Addendum 2: The reviewer for the publication in the journal *Distinction* has referred me to the preface by Spencer-Brown for the *Limited Edition* of *Laws of Form* from 1994, which I had not heard of before. There Spencer-Brown sees in the ideas of "conditioned coproduction, called by me the calculus of indications" (Spencer-Brown 1994b, viii) his predecessor. He cites Sakyamuni, the name of Siddharta Gautama, the founder of Buddhism, as his source. He sees him as "the only other author who evidently discovered these laws" (Spencer-Brown 1994b, viii). In contrast to the "falseness of current scientific doctrine, what I call scientific duplicity: that appearance and reality are somehow different," he represents the idea of "triplicity." This leads him to formulations that come very close to those of Nagarjuna: "there can be no appearance that is not an awareness of appearance and, of course, no awareness that is not an appearance of awareness." (Spencer-Brown 1994b, vii). If I understand Spencer-Brown's preface for the German first edition of *Only Two can play This Game* correctly, also written in 1994, he became aware on September 7, 1970, through an enlightenment, that his thoughts and possibly he himself were a rebirth of Buddha's thinking (Spencer-Brown 1994a, 5). The relationship between the statements in the preface of 1994 and those of Nagarjuna and other philosophers such as Hegel, who also thought in terms of trinities, needs to be examined separately.

4 Theory of Evolution and Adjoint Functor Pairs: from Instruction to Selection

With the theory of evolution, the concluding step was achieved: Not only is the possibility considered of whether something can be processed and determined, or whether something has the ability to do or to think something, but the subject and the object unfold into *fields of possibility* from which a multiplicity of processes can emerge. The rigid path of deterministic instruction is expanded into a multiplicity of possibilities from which one is selected. Ellerman formalized this with the methods of category theory (Fig. 8).

Ellerman chooses his terms from pedagogy. Traditional pedagogy gives instructions: The teacher tells the student what to do and what to say. Students memorize the multiplication table and other arithmetical forms. They can comply with the teacher's instructions, but in most cases, they don't know what they are doing and give up. Then they can be replaced by calculators and made redundant. It is similar with any other kind of work: Instructions can be taken over by robots. Selection, on the other hand, means that the student learns to find his or her own

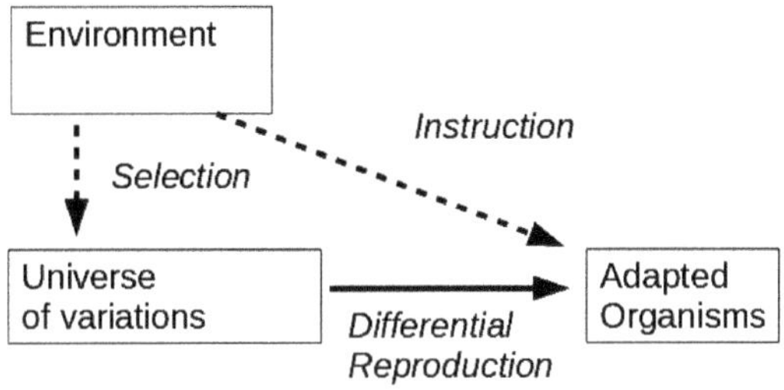

Figure 8: Theory of evolution. (Source: Ellerman 2007, 30, Fig. 8)

way of calculating from among an unknown variety of tasks and abilities. In my understanding, this corresponds to the power of judgment (*Urteilskraft*) meant by Kant and goes back to Aristotle's *Nicomachean Ethics.* Such selection does not contain prohibitions, commandments, and general instructions about how to act in a given situation but aspires rather to train the ability to judge and act independently.

Ellerman illustrates this with numerous examples. I choose the immune system as an example.

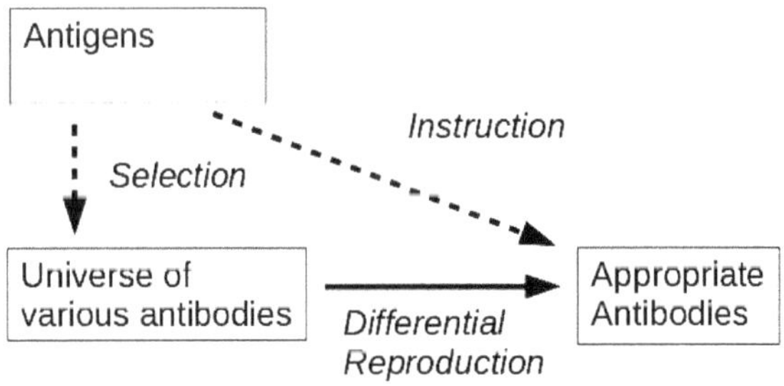

Figure 9: Strict and flexible immune response. (Source: Ellerman 2007, 32, Fig. 1)

The body can react immediately to some pathogens (antigens) with a prepared response. This corresponds to the instruction. Overall, however, it is impossible to be prepared for all potential dangers with an exact response. Therefore, the immune system holds prepared responses that are still unfinished, and these are assembled and activated only at the moment of a specific actualized danger (Fig. 9).

Addendum: One participant points out: When the active and the passive (e.g. the formable and the forming-capable, or the counting and the counting-capable) interact and awaken and inspire each other, there is not only a one-way direction toward a result, but at the same time a reverse direction. This becomes particularly clear in the example of the immune system. If the immune system detects the weak points of an antigen and develops corresponding antibodies, the antigen also learns the other way round and forms new variants with which the defense is to be circumvented. Recent examples of the COVID 19 virus mutations illustrate the point. As a result, there is not only a linear movement from a starting point to a result, shown here with the dashed diagonal arrows, but a second-order interaction that allows the relationships shown here to unfold spirally into a third level.

5 Spencer-Brown

In conclusion, this can be applied almost schematically to Spencer-Brown's basic concepts.

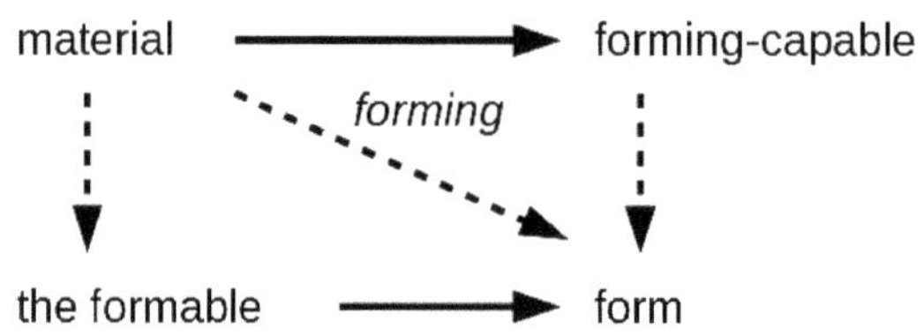

Figure 10: From a material to a form

The formable and the forming-capable are changed and shape a material into a form in two ways (leftward and rightward; Fig. 10). Also the observer is no longer externally facing the thing he observes, but is in an interaction with it (Fig. 11).

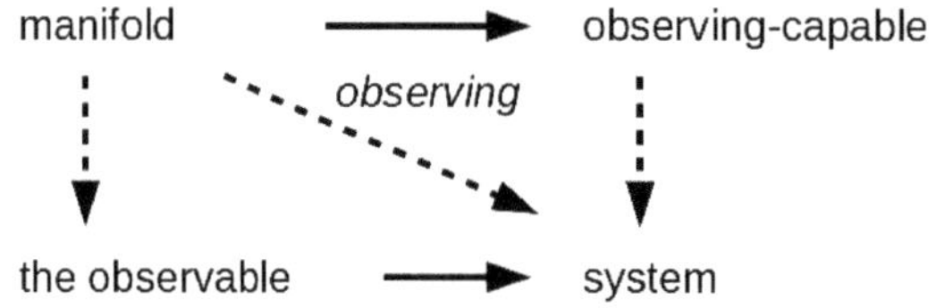

Figure 11: Observer

Spencer-Brown suggested this when he introduced a re-entry of the second kind for the observer. The one who has read *Laws of Form* and observed its content will subsequently understand himself better and could write the work anew and in an extended form.

Both the observed object and the observer change, *transforming an indeterminate manifold into a system.*

What is decisive is the interrelation of the two possibility fields. When they find each other, it leads to flourishing. Every rigid system includes sides thanks to which it can be set in motion (dynamized), and the ability to set something in motion is awakened when it meets the suitable material (Fig. 12).

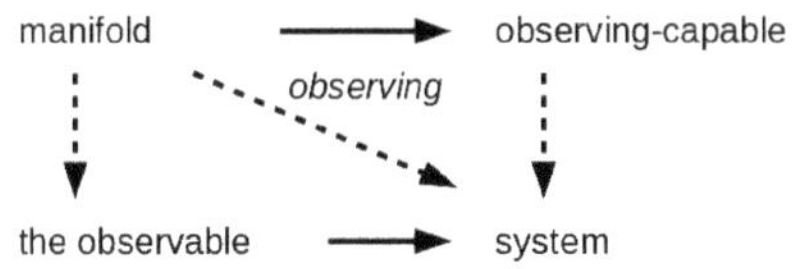

Figure 12: Flourishing

As a sign of flourishing, I choose the symbol of goddess Vesta, who comes from Roman mythology and who lights the *sacred fire* of the domestic *hearth* and keeps it alive: ⚶.

6 From a Paving Stone via a Path Sign to an Arrow

Finally, an example is given which goes back to the prehistory of category theory. Category theory operates predominantly with arrows and no longer with numbers. This is preceded by a development first from objects to numbers and then from numbers to arrows. One of the earliest documented examples of the use of an arrow is a waymark in ancient Ephesus, on which a symbolized footprint is imprinted as an arrow on a pavement stone.

Robert Finkel took this image from Flickr and interpreted it in his post "History of the Arrow":

Figure 13: Directions to a brothel drawn in the pavement of a street, Ephesus, 1st century before Christ. (Source: wikimedia.org)

> One of the earliest evidences of an instructional illustration is that of a footprint next to a woman's face. This pictograph is inscribed into the pavement of the ancient Greek city of Ephesus (now present-day Turkey) around the first-century AD. It is a reductive set of directions to the local brothel. The two symbols, a footprint and the woman's face, when united is read to suggest, 'Walk in the direction that the foot is pointing towards to reach the brothel.'

The footprint is prefigural: At the toes and the tapering of the foot with the middle toe and the big toe, the precursor of the arrow ↗ can be seen. And it is processual: It triggers the process of finding the way to a brothel. The processuality is the characteristic that has led to the formation of the symbol of the arrow.

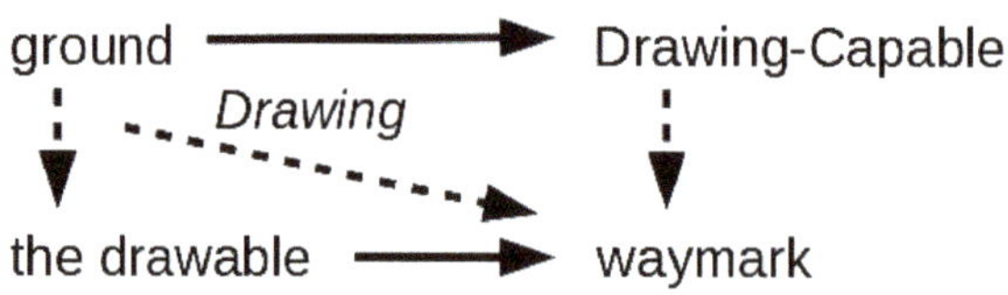

Figure 14: Drawing and Symbolizing

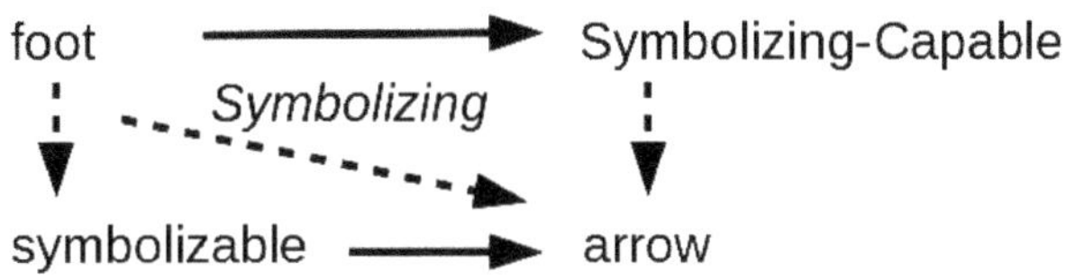

Figure 15: Drawing and Symbolizing (2)

7 The Flourishing of Identity in Storytelling

Leon Conrad asks if the arrow and the example from Ephesus are a story. Those who see this sign add to a story for themselves what those who followed this sign experienced.

In his article *The Sense of Sentences*, he asks, "What if there is more about sentences than we are taught at school?" A story is far more than a sequence of propositional sentences or even a form, such as the one given in the example of the waymark at Ephesus. It takes the listener out of his usual environment and brings him into the situation in which the story is set. It touches his own memories and expectations (in Husserl's terminology: his intentions, retentions, and protentions). In his imagination, comparable to daydreaming, he transcends the usual form (syntax) of ordinary language and enriches the words of the story with a wealth of unspoken meanings and feelings.

For me, therefore, the examples of the countable/counting-capable etc, mentioned so far should be extended in the direction of the nameable/naming-capable and the tellable/telling-capable. If we ask what is narratable, these are not merely the countable and measurable quantities to which conventional scientific research deliberately limits itself. Every story begins with names, such as: *Once upon a time there was a little girl, Snow White,* who lived behind *the seven mountains*, ... or imagine what happened to *Sinbad the Sailor* on *Monkey Island* ... Every name stimulates the imagination: What can the little girl or Snow White experience? What experiences do I want her to have, and what do I want to protect her from? With each proper name, the listener is directly involved and imagines the world (the field of possibilities), what will happen and how he could witness it and intervene by himself. This creates the excitement that a good narrator is able to generate. The narrator sees and senses how the listener is involved and reacts. A good narrator will immediately respond to this. He will never tell the same story twice in the same way, but each time in a new way, even for himself. Basically, he would have to rewrite the story each time. Every story has the inner power to stimulate the listener and inspire their imagination, as well as the storyteller themselves. The narrative will not appear to him as a fixed, static text, but as a field of possibilities in which he discovers a new, previously unseen variant with each new performance. With each new narration, he will allow himself to be carried away anew by the energy of the narration, the storytelling and the attentive listeners. (I am also writing this as a thank you to Leon, whose storytelling talent has delighted me every time).

This has far-reaching philosophical significance: A person's identity is not a static concept (profile) with facts such as birthday, place of birth, nationality, age, marital status, profession, etc, but a narrative that flourishes from the interdependent process of hearing and telling, being capable of hearing and being capable of telling (Fig. 16). This begins at birth: When parents give their child a name, they connect it with a narrative about the hopes and wishes they give the child, who this child wants to be, and what it wants to experience. Furthermore, everyone forms a narrative of how they see themselves and others in their lives. These narratives are changed and updated with every new experience. A person's identity lies in the narrative of how they see themselves and how others see them. This identity will outlast death.

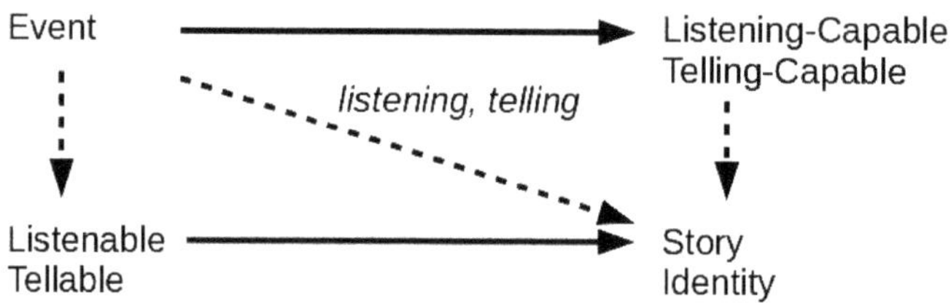

Figure 16: Storytelling

In the co-arising of telling and listening, intersubjectivity is realized. Only of that which has been heard before can be a tale be told. Every story is addressed to listeners, whether it be a soliloquy, a meditation or a prayer to possible listeners, whose presence is experienced almost bodily. Every storyteller will experience how his narration changes him no less than his listeners, thanks to the reactions of others. There is a second-order storytelling of what happens in storytelling. This goes far beyond the pragmatics of working contexts, from which the emergence of language is usually supposed to be explained.

8 The Flourishing of Imaginary Number Manifolds from the Impossible

In her article *Laws of Form in Husserl's 'Strange World of the Purely Logical'*, Claire Ortiz Hill referred to Husserl's remarks on imaginary numbers. By imaginary numbers, Husserl means not only the imaginary number i (the square root of -1), as in conventional mathematics, but all non-natural numbers such as zero, the negative, fractional and, in a stricter sense, imaginary numbers. These are numbers that do not result from counting or measuring material objects. There is nothing countable or measurable with a negative or imaginary number of elements. There is no measurable quantity that is smaller than 0 or that is imaginary. With fractional numbers it is at least possible to understand them as ratios $a{:}b$ of two natural numbers a and b, e.g. 3:4 or 1:1000. However, it is not the *ratio* that can be counted or measured, but only the two quantities a and b whose *ratio* is being considered. There is no three-quarter or thousandth ratio of two numbers; these would rather be ratios to be recalculated.

Anyone who still plays with numbers like a child with objects understands this immediately. If a child has learned to cut a piece of fruit and is told to break a number, they will laugh: "How should I hold the number in my hand, and where should I start?" Broken numbers only exist in the imagination; they are imaginary. Like Spencer-Brown, Husserl is trying to find his way back to an original approach to numbers. We are so far away from this through education and habituation that it becomes difficult to understand Husserl's concerns and to find the right words to explain them.

In mathematics and arithmetic lessons, there is an unspoken leap in method at this point. Whereas arithmetic calculation initially referred to given, countable things such as the fingers on a hand, apples or building blocks that lie in front of me and can be counted, geometrically arranged, or sometimes put together to form buildings, we now operate with purely mathematically defined objects that can no longer be counted or measured in themselves. They are not visible or otherwise observable, and certainly nothing can be built from zero or negative numbers. There are only jokes that achieve the impossible: "There are 5 people on a bus. At the next stop, 7 people get off. How many people have to get back on for the bus to be empty?" (bbq-treff.de) Such numbers are in the realm of the unobservable. If it is nevertheless possible to calculate with them and produce results that are themselves countable and measurable, i.e., have found their way back into the realm of the observable (re-entry), Husserl describes this as the *Transition through the Impossible (Durchgang durch das Unmögliche)*—the title of his programmatic double lecture in 1901 at the Göttingen Mathematical Society.

Mathematics leaves the realm of the clearly explainable. Teachers cannot explain it, and students can only memorize it. And yet these numbers are somehow given. For Husserl, this is the question of givenness: When is something considered given, and when is it only claimed to be given? How can we check whether it is really given? Do things with a negative count or size only exist in a fantasy world, as we know it, for example, from Lewis Carroll's *Through the Looking-Glass*?

The previous examples of independent co-arising need to be extended once again. While the relationships in counting and measuring were clear (there is something that can be counted or measured, and there is something that can count or measure it), things became more complex in narrating and hearing. In narration, pairs of the active and passive appeared on both sides of the tellable/listenable and the telling-capable/listening-capable: What is narratable must also be audible, and whoever can narrate must also be able to hear (and vice versa: what is audible must be narratable, and whoever can hear must also be able to narrate, at least in order to report back to the narrator whether and how he has understood him).

According to Husserl, the relationship between telling and hearing is generalized to the relationship between giving and taking and the variety of possibilities (the field of possibilities) of giving and taking, among which Husserl will distinguish three classes, each with its own logic. Telling can be understood as the giving of messages, and hearing as their taking. (The German words 'vernehmen' and 'aufnehmen' contain the root 'nehmen'. This cannot be translated literally.) The terms 'giving' and 'taking' provide the framework for understanding what Husserl is concerned with. For every given there is a variety of givennesses. Husserl initially thinks of very simple examples: A house is given differently whether it is seen from the front or the back, by the neighbors or by participants at a garden party. It is given differently when its architectural floor plan is viewed, or when potential buyers or tenants inspect the house to see whether it is comfortable for them to live in. 'Giving' can mean *giving* a thing its form. A thing *gives* itself to be recognized. For every *giving* there is given a multitude of givenesses.

There are not only sensorially perceptible things, such as the pebbles (*calculi*) with which counting and arithmetic were learned (from which the term 'calculus' is derived), of which Husserl writes: "In the sense of the stricter 'sensory' perception, an object is directly grasped or is itself present, which is constituted in a *straightforward* way in the act of perception." ("Im Sinn der engeren 'sinnlichen' Wahrnehmung ist ein Gegenstand direkt erfaßt oder selbst gegenwärtig, der sich im Wahrnehmungsakte in *schlichter* Weise konstituiert." Husserl 1921, § 46, 145, my translation using DeepL). Dreams, works of art, thoughts and overarching entities such as water (*hydor*) in continuous motion in the sense of Thales and Heraclitus are to be distinguished from this. Husserl refers to them as "the categorial or ideal as the objects of the higher levels" (Husserl 1921, § 46, 145, my translation using DeepL) and goes on to give his own examples: "*Color, house, judgment, desire* are purely sensory concepts; *colorfulness* (being colored), *virtue, parallel axiom*, and the like are categorially mixed; unity, majority, relation, concept are purely categorial." (Husserl 1921, § 60, 184, my translation using DeepL). Categorial or ideal objects can neither be counted nor measured. What is the number or size of dreams or of water as a whole? What makes a *great* work of art? But they can be calculated on! That is the special feature of imaginary numbers, whose ontological status has not yet been clarified by mathematics. If we succeed in understanding how to calculate with them and, above all, how imaginary numbers return from the realm of the impossible to the real world, then these numbers will make it possible to develop calculation methods for the other objects of the higher levels and to use them to 'recalculate,' as it were, how fantasy moves away from reality and returns to it. Husserl drafts the idea of a completely new understanding of mathematics.

On the other hand, there is a variety of ways of taking and hearing, for example when I hear the wishes and concerns of another person and understand them intuitively and compassionately without reducing them to specific objects, or understand what a composer and interpreter of a musical work wants to say to me without being able to find words for it. The givenesses on the one hand correspond to the intentions, memories and expectations (retentions and protentions) on the other. This gives for me an overview of some of Husserl's basic concepts:

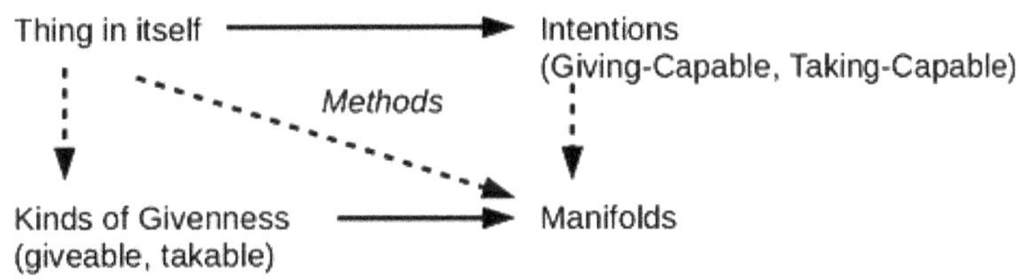

Figure 17: The kinds of givenness and the kinds of being capable of giving and taking

The example of negative numbers shows the direction this takes: In thought, it is possible to oversee processes and their inner movement and contradictions. There is nothing whose number or size is negative, but there are balancing processes. When a lever arm rises on one side, it sinks on the other. When someone lends money to someone else, he receives a document that he can redeem later with the repayment. Conversely, the borrower (creditor) has a debt on his account. These processes are the only way to understand what is meant by negative numbers such as a negative account balance (see Kant in his 1763 *Attempt to Introduce the Concept of Negative Quantities into Philosophy*). Like the axiom of parallels mentioned by Husserl, balancing processes are an example of 'objects of the higher levels.'

Without deepening or expanding on these examples, Husserl's insight is that each kind of givenness and the corresponding kind of taking-and-hearing-capable determines the giving and taking specific to it. The two fields of possibility of giving and taking lead to a third field of possibility of methods. There is not one method for giving and taking, comparable to the method of counting in the relationship between the countable and the counting capable, but rather a variety of methods. Within this diversity, balancing processes can occur between the different methods. Self-referentially, the methods can be used to infer the underlying forms of theories (Theorien-formen) in order to explain the safety, comprehensibility, applicability and simplicity of the associated method. Husserl did not work this out further, but saw this as the highest task of philosophy. The questions about the transition from ordinary to imaginary numbers and back from imaginary numbers to ordinary numbers (re-entry) can only be clarified with the form of theories.

> On the basis of a sufficiently developed solution of the given tasks, it becomes possible to formulate multiple concepts of possible theories from purely categorial concepts, pure 'forms' of theories, whose essentiality has been lawfully determined. However, these different forms are not without relation to each other. There will be a certain procedural order, according to which we will be able to construct the possible forms, to survey their lawful connections, and thus also to transform the one into the other through variation of determining basic factors, etc. There will be, if not in general, at least for the forms of theories, general propositions which, within the defined scope, will govern the lawful development of the forms, their interconnection and transformation. (Husserl 1913b, § 69, 247, my translation using DeepL)

> This is the ultimate and highest aim of a theoretical science of theory in general. It is also not an indifferent one from an epistemological-practical point of view. Rather, the classification of a theory into its form class can be of the greatest methodological importance. For with the expansion of the deductive and theoretical sphere, the free vitality of theoretical research also grows, the richness and fruitfulness of the methods increases. (Husserl 1913b, § 69, 248, my translation using DeepL)

These three sides mutually awaken each other in the flourishing of the cognition of a manifold/form of theories, in which Husserl distinguishes three stages of logic:

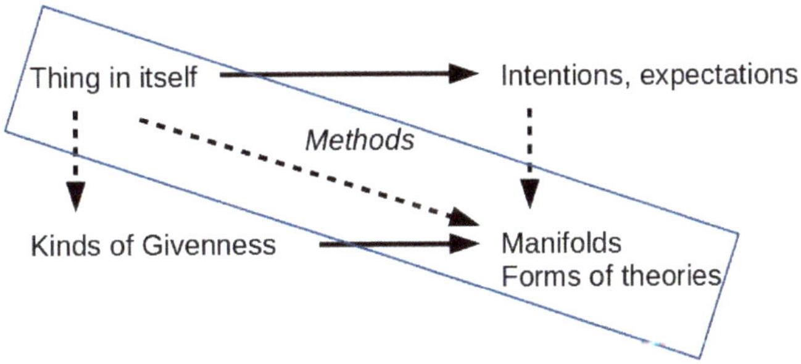

Figure 18: Three levels of logic according to Husserl

For Husserl, the forms of theories reach the stage at which manifolds of mathematical objects can be determined and freely operated with. The impossible numbers did not simply emerge from more or less arbitrarily defined calculations, but from the forms of theories that Husserl connected with the ordinary objects in his design and explained from them. For me, this development can be described as a flourishing of the forms that are already given with the ordinary objects and lead far beyond them. If we think of the diversity of categorial objects from dreams to axiom systems and the like, this flourishing can be associated with the revelation of the divine, just as Hegel saw it for the course of the science of logic.

In her contribution *Husserl's 'Mannigfaltigkeitslehre'* from 2000, Claire Ortiz Hill refers to a passage in Husserl's 1917/18 lecture *Logic and General Theory of Science*:

> The circumstance then that, instead of in the specific fields of knowledge, one can operate rather in the spheres of pure forms, and,

> if they are complete, can operate freely, then leads to a systematic expansion of the specific discipline-forms there and to pursuing all possibilities of the construction of specific discipline forms. (Husserl 1917/18, § 57, 265).

In her own words: "One may then, he concluded, operate freely within a multiplicity with imaginary concepts and be sure what one deduces is correct when the axiomatic system completely and unequivocally determines the body of all the configurations possible in a domain by a purely analytical procedure." (Hill 2000, 175)

Husserl oriented himself on the point manifolds introduced by Cantor, and conversely, Cantor's work on the power and order types of the number classes can be re-read from Husserl's findings. We must ask about the power (the energy) of form, which leads to its flourishing both on the side of the givenness and on the side of the intentions and the forms of theories. Cantor named them with letters of the Hebrew alphabet, studied their occult (hidden) properties and presumably saw in them the stages of infinity to the divine. A planned work will deal with this.

The result is open. As Claire Ortiz Hill has pointed out in other books with evocative titles such as *The Road not Taken*, in the 20th century the roads and directions going back to Frege and Husserl became separated and alienated from each other. Through his relationships with Wittgenstein and Russell, Spencer-Brown was on the whole on their path and probably neither knew Husserl, nor was he made aware of him from his friends and colleagues. Claire Ortiz Hill's contribution provides an initial impetus as to how the two can learn from each other and—albeit belatedly—find their way onto *the road not taken*. Anyone who wants to continue working on the topics mentioned by Husserl will find inspiration in Spencer-Brown, especially when it comes to understanding what is meant by form. Husserl sees form as the result of abstraction and formalization until its "pure form" is achieved (see e.g. Husserl 1929, §§ 6, 26). With Spencer-Brown, we can learn how form, for its part, is a content to be studied in its own right, with its own laws and developments. In the other direction, further research on Spencer-Brown's work with Husserl can find a new understanding of its basic assumptions and, for example, no longer simply say, "We take as given the idea of distinction and the idea of indication" (Spencer-Brown 1969, 1), but with Husserl ask about the modes of givenness and recognize which implicit restrictions for givenness have been presupposed and can be extended.

References

[1] Arnold, Vladimir I. 1989. *Mathematical Methods of Classical Mechanics.* 2nd ed. Translated by Kurt Vogtmann and Arnold Weinstein. Graduate Texts in Mathematics, vol. 60. New York: Springer. https://doi.org/10.1007/978-1-4757-2063-1.

[2] Baecker, Dirk. 2021. *Katjekte.* Leipzig: Merve.

[3] Conrad, Leon. 2025. "On the sense of sentences". This volume.

[4] Ellerman, David. 2007. "Adjoints and Emergence: Applications of a New Theory of Adjoint Functors." *Axiomathes* 17: 19–39. https://doi.org/10.1007/s10516-007-9008-x.

[5] Ellerman, David. 2015. "On Self-Predicative Universals in Category Theory." arXiv:1405.3192 [math.GM]. Submitted May 9, 2014 (v.1), last revised May 11, 2015 (v.2). http://dx.doi.org/10.2139/ssrn.2605120.

[6] Finkel, Robert J. 2015. "History of the Arrow." American Printing History Association. Posted April 1, 2015. https://printinghistory.org/arrow/.

[7] Hill, Claire O., and Jairo J. da Silva. 2013. *The Road Not Taken: On Husserl's Philosophy of Logic and Mathematics.* London: College Publications.

[8] Hill, Claire O. 2000. "Husserl's Mannigfaltigkeitslehre." In *Husserl or Frege? Meaning, Objectivity, and Mathematics*, edited by Claire O. Hill, Guillermo Haddock, and E. Rosado, 161–178. La Salle, IL: Open Court.

[9] Hill, Claire O. 2022. "Spencer-Brown's *Laws of Form* and Husserl's 'Strange World of the Purely Logical.'" Paper presented at the *Laws of Form 2022* Conference, University of Liverpool, August 3–7.

[10] Husserl, Edmund. 1901. "Durchgang durch das Unmögliche." In *Husserls Manuskripte zu seinem Göttinger Doppelvortrag von 1901*, edited by Karl Schuhmann and Elisabeth Schuhmann, 87–123. *Husserl Studies* 17.

[11] Husserl, Edmund. 1913a. *Ideen zu einer reinen Phänomenologie und phänomenologischen Philosophie.* Halle a.d. Saale.

[12] Husserl, Edmund. 1913b. *Logische Untersuchungen, Erster Band: Prolegomena zur reinen Logik*, 2nd rev. ed. Halle a.d. Saale: Max Niemeyer.

[13] Husserl, Edmund. 2019. *Logic and General Theory of Science: Lectures 1917/18 with Supplementary Texts from the First Version of 1910/11.* Translated by Claire O. Hill. *Husserliana: Edmund Husserl—Collected Works* 15. Cham, Switzerland: Springer.

[14] Husserl, Edmund. 1921. *Logische Untersuchungen, Zweiter Band: Elemente einer phänomenologischen Aufklärung der Erkenntnis, II. Teil*, 2nd, partly revised ed. Halle a.d. Saale: Max Niemeyer.

[15] Jeck, Udo R. 1994. *Aristoteles contra Augustinum*. Amsterdam: John Benjamins.

[16] McCulloch, Warren. 1965. "What is a Number, that a Man May Know It, and a Man, that He May Know a Number?" In *Embodiments of Mind*, 1–18. Cambridge, MA: MIT Press.

[17] Spencer-Brown, George. 2011. *Laws of Form*. Leipzig: Bohmeier Verlag. First published 1969 by George Allen & Unwin.

[18] Spencer-Brown, George [James Keys, pseud.]. 1971. *Only Two Can Play This Game*. Cambridge: Cat Books.

[19] Weber-Brosamer, Bernhard, and Dieter M. Back. 1997. *Die Philosophie der Leere: Nāgārjuna's Mūlamadhyamaka-Kārikās*. Wiesbaden: Harrassowitz.

[20] Weber-Brosamer, Bernhard, and Dieter M. Back. 2005. *Die Philosophie der Leere: Nāgārjunas Mūlamadhyamaka-Kārikās Übersetzung des buddhistischen Basistextes mit kommentierenden Einführungen*. 2. durchgesehene Auflage. Beiträge zur Indologie 28. Wiesbaden: Harrassowitz Verlag.

Distinct but Non-Exclusive Categories: Introducing the *n*-category-lemma for second-order observation

Steven Watson
University of Cambridge

Abstract

This paper presents the *n*-category-lemma, an extension of the tetralemma from Buddhist philosophy, as a framework for understanding complex systems beyond binary logic. It explores the distinct but non-exclusive nature of categories or entities, highlighting their interactivity and mutual influence. The framework examines the exponential scaling of potential interactions and exclusions among categories, providing a tool for comprehensive analysis in nonlinear, multifaceted systems. The article draws on Spencer-Brown's *Laws of Form* and Luhmann's general theory of autopoiesis for philosophical grounding, emphasizing the importance of distinction in understanding complex systems. Additionally, the article illustrates the application of this approach using Large Language Model generative AI in theoretical analysis, demonstrating its utility in discerning intricate relationships in systems composed of multiple interrelated categories.

1 Introduction

In the following, I introduce the idea of the n-category-lemma approach to second-order observation. This approach is directly informed by *Laws of Form* (Spencer-Brown 1969) and autopoiesis (Luhmann 1995). The assumption underpinning this approach is that objects, entities, categories, or systems have the quality of being distinct but non-exclusive which reflects the ontology in Spencer-Brown's notion of distinction and re-entry.

The n-category-lemma approach extends the idea of the *tetralemma* (also known as the four corners argument) in Buddhist philosophy from the 2nd century CE (Westerhoff 2009). The tetralemma can be thought of as an extension of two-valued logic (where a statement is either true or false) to include four possibilities: true, false, both true and false, and neither true nor false (Roth 2019). The n-category-lemma extends this to n categories. It is a tool for thinking about complex problems when binary logic is insufficient.

Central to this is the idea of *distinct but non-exclusive categorisation.* This posits that while categories are unique in their properties and definitions, the distinctiveness of the categories is a consequence of the interactions of the categories. Interaction is characterized by a mutual or reciprocal influence among categories. This implies that the relationships between categories are not static.

The n-category-lemma framework provides a structured representation of interactions among n features, categories or systems by considering what happens in a 2^n field of interaction and exclusion between categories. The total number of potential interactions scales exponentially with the number of categories. This scaling reflects the combinatorial increase in possible interactions and exclusions as more categories are considered. For instance, with three categories, there are 2^3 or 8 possible interaction spaces, each representing a unique combination of interactions and exclusions across the categories. This model supports the analysis of complex systems where multiple factors interact in non-linear and multifaceted ways.

Second-order observation means observing not only entities, phenomena or systems but also observing how these are observed. This perspective acknowledges the interconnectedness and mutual influences between these elements. Unlike first-order observation, which focuses on direct observation of phenomena, second-order observation reflects on the observing process itself. It considers how different observers might perceive and interact. In second-order observation, the process of observing itself implies that distinctions made are distinct but not exclusive. This means that while categories or phenomena are recognized as different, they are not seen as entirely separate and isolated. Instead, there's an understanding that these distinctions are interconnected and influenced by the act of observation itself (Von Foerster 2003; Luhmann 2000).

This is explained and developed in what follows, in which I draw on the approaches of Spencer-Brown and Luhmann as a philosophical basis for the approach. Spencer-Brown's *Laws of Form* provides a means of visualising the conceptual basis of the tetralemma and *n*-category-lemma, while Luhmann's de-ontologised and de-temporalized generalisation of autopoiesis provides a conceptualisation of the notion of distinct but non-exclusivity (Luhmann 1995). I begin with a discussion of *Laws of Form* and autopoiesis. I then introduce the *n*-category-lemma and then illustrate how it works practically using a Large Language Model generative AI (LLM). This is followed by further discussion on the compatibility and similarity of the *n*-category-lemma to the architecture of LLMs.

The *n*-category-lemma appears to have many possibilities in research and data analytics. But it is in a crude form at present and remains speculative. However, I hope in the following to have demonstrated rigour in terms of reporting the underlying philosophy.

Laws of Form and autopoiesis

The act of distinction is the primary and initial act in Spencer-Brown's calculus of indications (Spencer-Brown 1969). It is the process of drawing a boundary to separate a space into two distinct parts: the inside and the outside.

> *Distinction is perfect continence.*
>
> That is to say, a distinction is drawn by arranging a boundary with separate sides so that a point on one side cannot reach the other side without crossing the boundary. For example, in a plane space a circle draws a distinction.
>
> Once a distinction is drawn, the spaces, states, or contents on each side of the boundary, being distinct, can be indicated.

> There can be no distinction without motive, and there can be no motive unless contents are seen to differ in value.
>
> If a content is of value, a name can be taken to indicate this value.
>
> Thus the calling of the name can be identified with the value of the content. (Spencer-Brown 1969, 1)

The act of drawing a distinction involves creating a boundary that delineates one area or concept from another, ensuring that each side is contained and separate from the other. This separation is critical as it allows for the classification and evaluation of different states or contents based on their intrinsic or assigned values. When a boundary is established—like a circle on a plane defining inside and outside—it facilitates the recognition and valuation of differences, motivating distinctions. This motivation is rooted in the differing values perceived between the contents of the distinct areas. Once a value is recognized, it can be named, linking the name directly to the value of the content it represents. Thus, in Spencer-Brown's framework, the process of naming is not merely nominal but is an act of identifying and valuing the content within a clearly defined boundary.

Spencer-Brown acknowledges that the distinction is distinct but not exclusive; it is at the same time distinct from the unmarked space but is also in unity with it. When a boundary is drawn, it not only defines what is inside but also implies the existence of an outside. This act of distinction is not just a spatial (or temporal) separation but also a conceptual one. It is through making distinctions that we can identify, characterize, categorize, and communicate different aspects of our experience. As Spencer-Brown explains:

> Thus we cannot escape the fact that the world we know is constructed in order (and thus in such a way as to be able) to see itself.
>
> This is indeed amazing.
>
> Not so much in view of what it sees, although this may appear fantastic enough, but in respect of the fact that it *can* see at all (Spencer-Brown 1969, 105).

This reflection on the nature of perception and reality presents an understanding of perception as not merely biological or psychological process but an existential and philosophical one. By recognizing that the capacity to perceive is as significant as what is perceived, Spencer-Brown invites us to consider the broader implications of how we construct reality and our place within it—how our cognitive and perceptual frameworks shape not just our understanding of the world but our understanding of ourselves within that world—as an interplay between observer and observed, between knower and known.

Thus, the act of distinguishing something inevitably involves the thing itself in a loop of reference, i.e. self-reference. Spencer-Brown refers to this as re-entry. Re-entry is where a form (or distinction) is self-referential. Before any distinction is made, the state is indeterminate, meaning it lacks structure, boundaries, or defined properties. When a distinction is observed, it is as a boundary that defines an inside (marked state) and an outside (unmarked state). This act does not eliminate the indeterminacy outside of the boundary; rather, it structures part of the indeterminate space. The marked state can host a multitude of further distinctions (forms within the form), but these are not predetermined by the initial act of marking.

In *Laws of Form*, the mark ⌉ is the fundamental unit of distinction. It creates a boundary that separates the marked state from the unmarked state. The mark serves two purposes: it creates a distinction, and it becomes a name for the marked state. To refer to or indicate (i.e. observe) either state, one must inherently reference the mark that defines them. Re-entry is the act of crossing the mark again, effectively entering the space created by the initial distinction. This re-entry is not merely a return to a previous state; it is an act of indication that reaffirms the distinction. By re-entering the marked state, we indicate it as distinct from the unmarked state. If re-entry did not occur, there would be no further reference to the marked state. Without this ongoing reference, the distinction between the marked and unmarked states would lose its significance. The two states would effectively merge back into an undifferentiated whole, as if the mark had never been drawn in the first place. Therefore, re-entry is not optional but a continuous necessity for maintaining a distinction. It is a constant process of self-reference that upholds the separation between the marked and unmarked states. Without this ongoing re-entry, distinctions would be ephemeral and fleeting, unable to persist.

In the first chapter of *Laws of Form*, The Form, Spencer-Brown begins by stating, "We take the form of distinction for the form." This implies a pre-distinction state, a void or an indeterminate space lacking any inherent structure or boundaries. The act of drawing a distinction is the first creative act. This does not eliminate the indeterminate state entirely; it merely carves out a defined region within it. Thus, the mark is inherently paradoxical. Observation of the distinction—whether to the inside, the outside, or the boundary itself—requires a reference back to this mark. This is the seed of re-entry. The act of crossing the boundary (re-entry) is not just about movement; it is an act of re-affirmation. Each re-entry reinforces the distinction, preventing the marked and unmarked states from collapsing back into indeterminacy. The mark itself, as a symbol of distinction, implicitly refers back to the act of creating the distinction. It is a self-referential symbol that embodies both the creation and the maintenance of the distinction.

By embracing, rather than avoiding, self-referential paradoxes, one can explore the nature of mathematical and logical systems more deeply, argues Spencer-Brown.

> Put as simply as I can make it, the resolution is as follows. All we have to show is that the self-referential paradoxes, discarded with the Theory of Types, are no worse than similar self-referential paradoxes, that are considered quite acceptable, in the ordinary theory of equations. (Spencer-Brown 2014, x).

This is consistent with Luhmann's conceptualisation of *autopoiesis*, which is unsurprising since Luhmann's thinking was influenced by *Laws of Form* (see Luhmann 2000). An autopoietic system is self-referentially operationally closed but thermodynamically open to its environment. Its difference/distinction from its environment acknowledges its unity with its environment. Like *Laws of Form*, it begins with self-referential paradox. Autopoiesis for Luhmann represents the possibility of distinct but non-exclusive categorisation.

The emphasis on categories in this paper denotes a consideration of orthodox approaches to conceptualising distinction wherein categorization is both a mathematical and ontological abstraction, suggesting that categories are distinct and exclusive by nature. This orthodoxy underpins much of classical logic and set theory, where entities are grouped based on shared properties, and these groups are clearly delineated from one another without overlap.

Distinction as re-entry and the recursion of autopoiesis provide a non-essentialist approach to comprehending categories as distinct but not exclusive. This offers an alternative approach to examining complex systems, emphasizing that the stability of categories is continuously achieved through dynamic processes, rather than being inherently fixed. For example, within ecological systems, species and ecosystems do not merely interact; they constantly adapt and evolve in response to these interactions. Such continuous adaptation can result in the emergence of new categories, like hybrid species or ecotones, challenging traditional taxonomic classifications.

Rather than primarily developing on his general theory of autopoiesis, Luhmann prioritised sociological analysis: social systems of communication and psychic systems of thought. His insight in this, however, emphasises that science, philosophy and mathematics are indeed social systems whose media are truth statements when examined not from an axiomatic starting point but from a foundation of self-referential paradox (Luhmann 1995).

Self-referential paradox implies that these systems both produce and are constituted by their own elements. For example, in science, truth statements (claims considered factual within the system) are both products of scientific activity and the basis on which the system operates. This circularity challenges the conventional notion of an external, objective foundation for these fields, suggesting instead that they are grounded in their own internal logic and practices.

Distinct and exclusive categories are fundamental in science, philosophy and mathematics. They underpin the veracity of truth claims by ensuring clarity and precision in observations and analyses. In these fields, clear categorization is critical for defining concepts, conducting analyses and drawing conclusions. The presence of non-exclusive categories, where elements belong to multiple categories, introduces ambiguity. This ambiguity can be likened to noise, diminishing the accuracy with which observations represent phenomena. Consequently, the potential to make definitive truth claims is weakened. Ambiguity due to overlapping categories erodes the necessary clarity for making such assertions, challenging the precision essential for rigorous analysis.

What both Spencer-Brown and Luhmann achieve is an approach to scientific analysis based on a non-axiomatic paradoxical referentiality. Luhmann applied the concept of self-referential paradox to social systems; Spencer-Brown applied it to the foundation of mathematical and logical systems, offering a new way to understand the formation and functioning of systems based on the simple but profound act of making distinctions. Spencer-Brown's methodology is arguably more abstract, focusing on logical and mathematical structures, while Luhmann's is more empirical, dealing with observations of social phenomena.

Spencer-Brown's approach to mathematics and logic, centred on the act of making a distinction, can be considered abductive because it begins with a fundamental operation (making a distinction) and explores its implications and the structures that arise from it. It does not start with a predefined set of axioms (deduction) or specific observations (induction). Luhmann's social systems theory can also be seen as abductive in nature. He starts with the observation of social systems and communication processes and develops explanatory frameworks for how these systems function and maintain themselves.

George Spencer-Brown's *Laws of Form* and Niklas Luhmann's general theory of autopoiesis represent a paradigm shift. Spencer-Brown's work transcends traditional logic by introducing a notation system that symbolizes both the presence and absence of a state, offering a more fluid interpretation of boundaries. This acknowledges that defining an entity involves recognizing not just what it is not, in other words, a resolved binary relationship between the thing and its negation. The mark is differentiated from infinite possibility but remains in unity with it. Infinite possibility in this regard can be understood at the same time as indeterminacy, complete absence and therefore void. In a state of infinite possibility, there are no constraints or limitations on what can exist or occur. In a situation where anything can happen, it is impossible to ascertain what will happen and it is therefore indeterminate. Complete absence can be interpreted as a state where nothing specific exists or occurs—since in a realm of infinite possibilities, no single possibility has been actualized.

In sum, the existence of anything is a differentiation from infinite indeterminacy/ void but at the same time in unity with such an unmarked state.

Spencer-Brown's idea of distinction is equivalent to the general notion of autopoiesis as presented by Luhmann. Autopoietic systems are simultaneously distinct from, but in unity with, their environment. He argued that social systems (like legal, economic or educational systems) are autopoietic. They maintain their existence through continuous self-reproduction, based on communication. Each system operates according to its own logic and is distinct from its environment, yet it cannot exist without interacting with that environment. For example, the legal system is distinct from the political system, but it interacts with it and is influenced by it.

The n-category-lemma for distinct but non-exclusive categories

As scientific knowledge expands and evolves, categories that were once considered stable and universally applicable are increasingly becoming inadequate. Gödel's theorems suggest that in any consistent formal system sufficient for number theory, there are true statements that cannot be proven from within the system. A system is incomplete without an environment, the implication of this is that to exist, a system must be autopoietic, i.e. self-referential and operationally closed. Similarly, quantum physics, particularly the concept of superposition, shows that particles exist in multiple states or categories simultaneously, until observed axiomatically. Advances in genomics have revealed the fluidity of gene exchange between species, suggesting that Linnaean taxonomic categories may not always reflect the true nature of evolutionary relationships.

The advent of advanced technologies like big data analytics and machine learning and deep learning further complicate traditional classification systems. These technologies uncover patterns and relationships that transcend conventional categories, revealing a more intricate tapestry of connections than previously recognized.

The n-category-lemma approach provides a structured approach to examining the structural relationships between n categories. This draws on the idea of second-order observation, *the observation of observation*, where the n-category-lemma is a means of considering how categories are formed, identified, and utilized within the system itself.

The principles of *Laws of Form* can be used to explain this. Let us begin with two systems or categories x and y. In the following they are independent categories or systems.

In *Laws of Form* notation: $\overline{x|}\,\overline{y|}$

Or in parenthetic form: $(x)(y)$.

The unwritten cross (Spencer-Brown 2014, 6) indicates the presence of the unmarked state, a state of being that is essential for the formation of any distinction yet remains invisible and implicit within the system. Observing or considering two systems as related invokes the presence of an 'unwritten cross' or mark ((x)(y)). When multiple distinctions (systems) are observed or considered in relation to each other, the 'unwritten cross' comes into play. The relationship itself (the interaction or connection between x and y) is an integral, though not explicitly stated, part of the systems' representation.

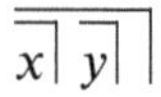

The above shows a *Laws of Form* representation of the ancient Buddhist and Jainian logical method, the tetralemma.

The tetralemma

In parenthetic form ((x)(y)) represents the logic of the tetralemma:

Neither x nor y (both x and y (x) (y))

In *Laws of Form* notation:

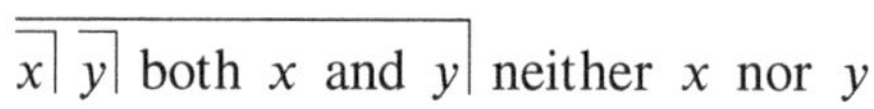

Therefore, ((x)(y)) in this context symbolizes a complex interplay of distinction and unity. It suggests that when considering multiple systems or categories in relation to each other, one must account for their individual identities (x and y), their potential interaction or coexistence (both x and y), and a transcendent state that is neither x nor y yet arises from their relation.

To exemplify this, consider the categories of human and machine. A tetralemma analysis would result in four categories. The first, exclusively human, encompasses attributes unique to humans, like emotional intelligence, creativity and the capacity for abstract thought, emphasizing the singular aspects of human consciousness and ethical reasoning. The second category—exclusively machine—includes features unique to machines, such as high-speed data processing and the capacity to perform large-scale computational tasks without fatigue, underscoring their strength in algorithmic operations and data management. This contrasts with the third category, which explores the interplay between human and machine, and leads into the fourth: a space beyond both, representing an indeterminate or unmarked domain from which all distinctions emerge.

It is important to recognise in this how these categories are defined in relation to other parts of the system rather than having essential properties. This is an important feature of this analysis. The third category, both human and machine, represents an emergent domain, where the interaction of human and machine have emergent properties. This means that the interactions are complex and not simply a summation of the properties of human and machine.

For example, in an educational setting, augmented reality technology can be used to create interactive learning experiences. The AR system (machine) provides a layer of digital information—such as images, data, or animations—over the real-world environment. When students (humans) interact with this augmented environment, they engage in a learning process that is qualitatively different from traditional or purely digital learning methods.

The collaborative process between a human and an AI system can result in decision-making or problem-solving approaches that neither the human nor the AI could have achieved alone. This could include enhanced creativity, improved analytical capabilities, or innovative solutions to complex problems. The emergent properties in this case are the result of the synergistic interplay between human intuition, experience, and creativity, and the machine's computational power, data processing capabilities, and algorithmic precision.

The 'neither human nor machine category' is the unmarked state. It is an indeterminate infinity of possibility from which the distinction of both human and machine are distinct, within which the categories of human and machine are defined. The existence of this category also highlights the importance of contextual and relational understanding in defining categories. It suggests that the understanding of 'human' and 'machine' is not just inherent within these entities but is also defined by what they are not. In other words, the identities of 'human' and 'machine' are not just self-contained but are also shaped by their relationship to this undefined, limitless category. It should be evident in this that distinct categories of human and machine are inherently founded on their relationship to each other. The unique identities and roles of humans and machines are not just defined in isolation, but also in how they relate and interact with each other.

This does not suggest that the distinctions (as categories) within social and psychic systems are being observed directly. In this, the observing system is observing the social system, i.e. communication. Communication within an autopoietic social system involves the operation of communication, where a communication operation is, in Spencer-Brown's terms, the drawing of a distinction. Each communication draws a distinction that differentiates information, thereby constructing the social reality (environment or unmarked state) within which the system operates. A communicative distinction might be a distinction within distinction(s) within communication.

Communicative distinctions between 'machine' and 'human,' imply the observation of a communicative operation. This is a first-order observation. The continued differentiation of machine and human implies that each of these distinctions re-enter respectively the distinctions of machine and human. They are continuously revisited, reaffirmed, and redefined through ongoing communication processes. This does not mean that the distinction between machine and human is the only one at play—rather, such distinctions emerge within communication itself. However, it can be understood as a relevant distinction in relation to the human-machine interface.

The tetralemma provides (as does *Laws of Form* and general autopoiesis) a means of observing the observation of the distinction between human and machine by assuming that they are re-entering distinctions or autopoietic. This second-order observation provides a theoretical second-order analysis of the observation of distinctions. This allows access to the complex interplay between distinctions as distinct but non-exclusive categories. This contrasts with the orthodoxy of distinct and exclusive categories, binary distinction, and essential entities.

Extending the tetralemma to n categories

The following represents a summary of my research into the n-category-lemma in the last three months of 2023. Initially, I present a formalised combinatorial set of instructions for constructing an n-category-lemma. This is less challenging than the second part of the analysis, which involves interpreting the vertices within the n-category-lemma.

The n-category-lemma extends the tetralemma analysis to n categories. This includes n distinct but non-exclusive categories or autopoietic systems; it represents possible interactions between all categories (subsystems within a system of distinctions) that include all combinatorial interactions and exclusions of the n categories, for example, the interaction of c_1, c_2, c_3 excluding, c_4, c_5, $\ldots c_n$, etc and, finally, the interaction of all exclusions, which represents the unwritten cross.

For an n-lemma, there are 2^n vertices,[1] each corresponding to a unique combination of included and excluded categories. Let v_j represent a vertex, where j ranges from 1 to 2^n.

To specify the characteristics of each vertex, two sets can be associated with v_j:

I_j is the set of categories included in v_j.

E_j is the set of categories excluded in v_j.

These sets are such that $I_j \cup E_j = C$ and $I_j \cap E_j = \emptyset$.

The formal definition of each vertex v_j can be captured as a mapping:

$$v_j \to (I_j, E_j)$$

For example, in a 4-lemma ($n = 4$), v_5 would map to:

$$v_5 \to (\{c_1, c_2\}, \{c_3, c_4\})$$

Mathematically, $V = P(C)$ and $|V| = 2^n$, where $|V|$ represents the cardinality of V and $P(C)$ represents the powerset[2] of C.

The following table shows the interactions of a 4-category-lemma with 2^n vertices.

[1]This reflects the four corners argument in earlier use of the tetralemma.

[2]In set theory, a powerset of a set S is the set of all possible subsets of S, including both the empty set and S itself. Mathematically, the powerset $P(S)$ is formally defined as:

$$P(S) = \{T : T \subseteq S\}$$

The cardinality (size) of the powerset $P(S)$ is 2^n, where n is the cardinality of the original set S. This formula holds because each element of S can either be present or absent in a

Vertex number	Categories included	Categories excluded
v1	c1	c2, c3, c4
v2	c2	c1, c3, c4
v3	c3	c1, c2, c4
v4	c4	c1, c2, c3
v5	c1, c2	c3, c4
v6	c1, c3	c2, c4
v7	c1, c4	c2, c3
v8	c2, c3	c1, c4
v9	c2, c4	c1, c3
v10	c3, c4	c1, c2
v11	c1, c2, c3	c4
v12	c1, c2, c4	c3
v13	c1, c3, c4	c2
v14	c2, c3, c4	c1
v15	c1, c2, c3, c4	None
v16	None	c1, c2, c3, c4

Each category c_i is distinct but not mutually exclusive with respect to other categories in C. The distinctness ensures that each category has unique characteristics that differentiate it from all other categories. On the other hand, the non-exclusivity implies that categories can co-occur or overlap in various ways.

subset, yielding 2 choices for each element. Since there are n elements in S, this results in 2^n possible subsets, which constitute the powerset $P(S)$.

In the context of an n-lemma with n categories $C = \{c_1, c_2, \dots, c_n\}$, each vertex v_j represents a unique combination of these categories. The set of vertices V is the powerset of C, $P(C)$, since each vertex is a subset of C. The architecture of the n-lemma, therefore, utilizes the powerset to account for the distinct but non-exclusive categories within the complex system or idea being modelled.

Interpreting the vertices of the n-category-lemma.

The 4-category-lemma in the above can be represented by *Laws of Form* notation in the following way.

$$\overline{\overline{c1}|\;\overline{c2}|\;\overline{c3}|\;\overline{c4}|\;}|$$

This represents v1 as c1, v2 as c2, v3 as c3 and v4 as c4, all understood as distinct but not exclusive categories. V15 is the interaction of c1, c2, c3 and c4, i.e., it is the space outside of the inner mark but inside the outer mark (unwritten cross). V16 is the unmarked space (outside the unmarked space).

V_5 comprises the inclusion of c1 and c2 and the exclusion of c3 and c4, and can be represented as:[3]

$$\overline{\overline{\overline{c1}|\;\overline{c2}|\;}|\;\overline{c3}|\;\overline{c4}|\;}|$$

The n-category-lemma proceeds iteratively with an initial dataset and/or initial categorization. This initial categorization assumes that the categories provide a complete representation of internal categorizations of the autopoietic system that the n-category-lemma represents. The initial categorization is a starting framework that attempts to capture and describe these elements and their interrelations as fully as possible. However, this is an iterative process, and the adequacy of this initial categorization may be tested and refined through further analysis and observation.

n-category-lemma theoretical analysis using large language model

While the approach shows promise in data analytics, my initial explorations have been theoretical, using a large language model (GPT-4). This involves considering

[3] This introduces a degree of hierarchical analysis, but needs further investigation.

categories theoretically. For example, we can observe a human-technology autopoietic system in terms of the categories human, society and technology as a 3-category-lemma with 8 vertices.

GPT-4 prompt:
The following are categories, c, for a framework for theorising the relationship between human, technology and society. These are to be treated as distinct but not exclusive categories. Do an n-category-lemma analysis which is a tetralemma but for n categories. For each vertex v, describe the emergent properties N=3: 1 human; 2 technology; 3 society. Include all of 2^n vertices starting with v1: c=1 human, to v7 the interaction of all categories and v8 the exclusion of all categories.

GPT-4 response:

*In an n-category-lemma analysis applied to the relationship between human, technology, and society, we consider a tetralemma framework extended to n categories, in this case, n = 3. This approach leads to 2^**n**, or 8 vertices, each representing a unique combination of the categories. Here is an analysis of each vertex:*

Vertex v1 (c=1: Human):
Emergent Properties: *This vertex focuses solely on human aspects, emphasizing human behavior, psychology, and sociology. It neglects technological and societal influences, thus isolating human nature and individualism.*

Vertex v2 (c=2: Technology):
Emergent Properties: *This vertex isolates technology, exploring its development, impact, and functionality independent of human or societal influence. It delves into aspects like innovation, automation, and digital transformation.*

Vertex v3 (c=3: Society):
Emergent Properties: *Here, the focus is on societal structures, cultural norms, and collective behaviors. This perspective excludes direct human and technological influences, concentrating on social systems, governance, and community dynamics.*

Vertex v4 (c=1+2: Human and Technology):
Emergent Properties: *This combination explores the interplay between humans and technology, such as human-computer interaction, technological adaptation, and the influence of technology on human behavior.*

Vertex v5 (c=1+3: Human and Society):
Emergent Properties: *This vertex examines the relationship between individuals and society, including social influence on human behavior, societal roles, and the impact of cultural norms on individuals.*

Vertex v6 (c=2+3: Technology and Society):
Emergent Properties: *This focuses on how technology influences and is influenced by society, including topics like the societal impact of technological advancements and the social shaping of technology.*

Vertex v7 (c=1+2+3: Human, Technology, and Society):
Emergent Properties: *This vertex represents the interaction of all categories. It encompasses the comprehensive analysis of how technology mediates human-society interactions, the societal adaptation of technology, and the co-evolution of these three elements.*

Vertex v8 (Exclusion of all categories):
Emergent Properties: *This vertex represents the absence of all three categories. It is a theoretical construct to understand what remains or emerges when human, technology, and society are all excluded, potentially leading to abstract concepts like void, non-existence, or purely natural phenomena untouched by human, technological, or societal influence.*

Each vertex in this analysis provides a distinct lens through which to view the complex interplay between humans, technology, and society, with the emergent properties shedding light on specific aspects of this triadic relationship.

Explanation

The above simple analysis illustrates the n-category-lemma approach using GPT-4. The prompt provided for analysis is a request to apply a theoretical framework to the study of the interconnections between human entities, technological entities, and societal structures. They are assumed to be autopoietic systems within an autopoietic system that integrates these distinct but non-exclusive categories. The n-category-lemma analysis considers n categories (in this case, three: human, technology, society) and explores all possible combinations of these categories. Emergent properties refer to characteristics or phenomena that arise from the interplay of categories that cannot be fully understood by considering any single category in isolation. In a system with three categories, emergent properties would be the outcomes or behaviours that result from the interactions between humans, technology and society. The introduction of the concept of the emergence emphasizes the systemic analysis. Each vertex provides an interpretation of category interaction based on GPT-4's training data.

Emergent properties in such a system might include forms of social interaction enabled by technology, alterations in societal norms due to technological adoption, or the evolution of human cognition in response to technological advancements. These properties are not solely attributable to any one of the three categories but rather to their dynamic interrelation. The n-category-lemma thus offers a methodological approach to discerning the complex and often non-linear relationships that underlie systems composed of multiple, interrelated categories. Niklas Luhmann's theory of social systems, particularly his concept of structural coupling, aligns with the n-category-lemma analysis in that it describes the interactions between distinct systems. In Luhmann's theory, structural couplings are the means by which systems stabilize their relationships with each other and their environment, enabling the exchange of information and co-evolution.

Within the context of the n-category-lemma, each vertex represents a potential *structural coupling*[4] (Luhmann 1995), where the included categories interact and stabilize their relationship, forming a complex web of interactions and dependencies. These couplings are emergent properties of the system—they cannot be deduced

[4]The n-category-lemma approach provides a clarification and elaboration of the concept of structural coupling in relation to autopoiesis. In this (and following *Laws of Form*), a structural coupling can be understood as the state within a distinction in which further distinctions are made. In the tetralemma a structural coupling between machine and human is denoted by the state 'both human and machine.' However, structural couplings exist relative to the observer. If the human-machine distinction is observed as different forms of meaning in the system of communication, meaning can also be comprehended as a structural coupling. The machine is given distinct meaning to the human. Thus, the n-category-lemma provides a means of exploring structural coupling from different observational perspectives, i.e. in different contexts. Orthodoxly, structural coupling refers to the relationship between two or more systems where each system influences and is influenced by the other while maintaining its own operational closure. The structural coupling between systems, i.e. distinction, is better understood as the distinction/system within which they are differentiated. Understanding structural coupling as the state within a distinction in which further distinctions are made emphasizes the dynamic and self-referential nature of structural coupling.

by looking at the individual components in isolation but are the result of their interaction.

For example, the coupling between the categories of 'human' and 'technology' may produce emergent properties such as augmented cognitive capabilities or shifts in social behaviour due to technology use. Similarly, the interaction of 'society' and 'technology' might lead to emergent properties like new forms of governance or collective intelligence.

In Luhmann's terms, these interactions would not be static but dynamic, constantly evolving as the systems themselves change. The n-category-lemma, like Luhmann's social systems theory, provides a framework for analysing these dynamic and complex relationships, offering a structured method to identify and describe the emergent properties that arise from structural couplings between different systems or categories.

Utilising Large Language Models in n-category-lemma analysis

The following represents an exploration of how Large Language Models (LLMs) like GPT-4, can be used to further develop n-category-lemma analysis. LLMs excel in detecting intricate language patterns, akin to the categorial interplays within the lemma. These models are adept at identifying linguistic nuances and overlaps, as they are trained on diverse textual datasets. They can discern subtle linguistic relationships, manage overlapping categories, and generate coherent, novel text outputs, which reflect the lemma's emergent properties. LLMs' systemic approach to language processing resonates with the multi-dimensional categorization of n-category analysis, enabling them to adapt and evolve understanding iteratively. Comparable to autopoietic systems, LLMs modify their parameters autonomously to refine their analytical capabilities in response to new or recontextualized data.[5] This

[5]This implies that technology is an autopoietic system. This follows a non-essentialist argument, that a distinction or system must be autopoietic otherwise it would have an essence. Following both *Laws of Form* and Luhmann's general theory of autopoiesis, the essence of any distinction or system lies in its distinction from an environment, but in paradoxical unity with it—its continued existence depending on recursion or re-entry. (For further discussion on this, see, Watson 2023, In review; Watson and Romic 2024.)

adaptability and self-updating nature make LLMs a potent tool for investigating the complexities of systems, enhancing our capacity to analyze and interpret the elaborate networks of relationships that traditional methods may not fully capture.

Fine-tuning or training an LLM on a specialized dataset or on theoretical perspectives can significantly enhance the sophistication of n-category-lemma analysis. By customizing the model with domain-specific data, such as texts on systemic theories, interdisciplinary research, or complex categorial frameworks, the LLM can develop a more nuanced understanding of the specific language, concepts, and patterns relevant to n-category analyses. This tailored training would enable the model to identify emergent properties and structural couplings. Moreover, incorporating theoretical perspectives into the training regimen could provide the model with a richer conceptual background, allowing for deeper insights and more precise analyses of the intricate relationships between categories.

The architecture of a Large Language Model (LLM) operates in a manner akin to n-category-lemma analysis, where multi-layered neural networks parallel the systemic interactions between categories. Neurons in an LLM's neural network are not dedicated to single, isolated linguistic features. Instead, they have the capacity to represent a range of features. This could include aspects like word meanings, parts of speech, grammatical structures, or contextual relationships. Each neuron can contribute to the representation of multiple features, akin to the way categories in n-category-lemma analysis are not exclusive to a single concept. Initial layers of an LLM recognize simple patterns, much like basic categorical interactions, and through a hierarchical process, akin to the lemma's structure, it synthesizes complex linguistic features from these patterns. This process captures the essence of non-exclusive, overlapping categories, with neurons mirroring this overlap while maintaining distinct activation patterns. The emergent linguistic phenomena produced by the LLM reflect the emergent properties of the lemma, derived from the network's collective processing rather than individual data points. Additionally, LLMs adapt and refine their internal parameters during training, much like the self-referential, autopoietic systems in n-category analysis, enhancing their capacity to accurately model the linguistic environment.

Conclusion

In conclusion, the n-category-lemma introduced in this paper contributes to understanding complex systems. Luhmann's theory acknowledges the presence of paradoxes as inherent to self-referential systems. He argues that these systems maintain their identity through continuous operations that both differentiate, and interact with, other systems. In other words, systems—and therefore categories—are fundamentally paradoxical from a non-essentialist perspective, and their forms are maintained through recursion or re-entry.

Transcending binary classifications enables a deeper, more holistic analysis of systems characterized by multiple, interacting categories. This framework, grounded in both Eastern philosophical principles and modern theoretical constructs, offers an approach to analyzing intricate relationships within and across diverse fields. Its application, as demonstrated in theoretical analysis using GPT-4, underscores its potential to enhance comprehension of multifaceted phenomena.

One of the limitations of analysing the complexity of autopoietic systems is the intensity of processing which places limits on technology. The n-category-lemma analysis can potentially make use of LLMs to carry out such analysis.

References

[1] Luhmann, Niklas. 1995. *Social Systems*. Translated by John Bednarz Jr. Stanford, CA: Stanford University Press.

[2] Luhmann, Niklas. 2000. *Art as a Social System*. Translated by Eva M. Knodt. Stanford, CA: Stanford University Press.

[3] Roth, Steffen. 2019. "Digital Transformation of Social Theory: A Research Update." *Technological Forecasting & Social Change* 146: 88–93. https://doi.org/10.1016/j.techfore.2019.05.016.

[4] Spencer-Brown, George. 1969. *Laws of Form*. London: George Allen & Unwin.

[5] Spencer-Brown, George. 2014. *Laws of Form*. 6th ed. Leipzig: Bohmeier Verlag.

[6] von Foerster, Heinz. 2003. *Understanding Understanding: Essays on Cybernetics and Cognition*. Edited by Albert Müller and Karl H. Müller. New York: Springer.

[7] Watson, Steven. 2023. "Technology as an Autopoietic System." *ResearchGate Preprint.* https://doi.org/10.13140/RG.2.2.14096.99846.

[8] Watson, Steven. In review. "Emergent Discourse on Generative AI in Education." Cambridge: Cambridge University Press. https://doi.org/10.13140/RG.2.2.29773.14562.

[9] Watson, Steven, and Jonathan Romic. 2024. "ChatGPT and the Entangled Evolution of Society, Education, and Technology: A Systems Theory Perspective." *European Educational Research Journal* (January). https://doi.org/10.1177/14749041231221266.

[10] Westerhoff, Jan. 2009. *Nāgārjuna's Madhyamaka: A Philosophical Introduction.* Oxford: Oxford University Press.

C. S. Peirce Meets George Spencer-Brown: Seeking semiotic self-evidence for logic

Jonathan J. Mize
Philosophy & Religion
University of North Texas
jonathanmize@my.unt.edu

To understand the foundation of mathematics it is necessary to abandon the symbolic representation of mathematics.—William Bricken

Abstract

Although there is ongoing debate about the mathematical influence of George Spencer-Brown's novel formalisms, one thing can hardly be denied: there is an intuitive and aesthetic simplicity to his notation which, if nothing else, deserves scholarly admiration and consideration. Of course, it is one thing to simply sit back and imbibe the simplicity of Spencer-Brown's "distinction" or "mark" and another thing entirely to meditate on what exactly makes these notions so intuitively pleasing and "amenable to common sense." I invite the reader to both indulge in the former and dabble in the latter, as we journey into an investigation of self-evidence as it relates to Spencer-Brown's notations and, moreover, logical notation in general. In this paper, I draw on C. S. Peirce's semiotics and philosophy of mathematics and Spencer-Brown's

¶ Paper based on presentation at the Laws of Form Conference 2022 in Liverpool, England.

passion for maximally self-evident formalism to craft a new breed of logical representation. I also highlight Hans Freudenthal's masterful "Lincos" an underappreciated language devised to be understood by any conceivable form of extraterrestrial life. In the end, I offer a style of self-evident logical notation that effectively collapses the boundary between (1) metalanguage and object language, (2) definiendum and definiens and (3) sign and universe of discourse.

1 Introduction

Cicero is rumored to have said that if the truth were self-evident, eloquence would be unnecessary. Any self-aware mathematician must admit something similar: If mathematical truth were self-evident, *proofs* would be unnecessary (and our mathematician friends would be sleeping on our couches!). Thankfully for mathematicians, mathematical truth is far from self-evident. Moreover, long gone are the days of Euclid, when *axioms* themselves adhered to some loose standard of self-evidence. As Bertrand Russell said, only half in jest, "The axiomatic method has many advantages over honest work."

Spencer-Brown himself of course recognized as much:

> Accounts of Boolean algebras have up to now been based on sets of postulates. We may take a postulate to be a statement which is accepted without evidence, because it belongs to a set of such statements from which it is possible to derive other statements which it happens to be *convenient to assume.* The chief characteristic that has always marked such statements has been *an almost total lack of any spontaneous appearance of truth.* [emphasis added] (Spencer-Brown 2011, xiv)

Spencer-Brown then goes on to say of his own system, "whatever our views on the nature of [the primary arithmetic's laws'] self-evidence, [they] at least recommend themselves to the findings of *common sense.*" [emphasis added] (Spencer-Brown 2011, xiv) Yet, although Spencer-Brown of course meant well—and went on to offer a rigorous proof of Sheffer's postulates with his system—the philosopher cannot help but inquire: *common sense?* If we still struggle to simply explain the notion of *self-evidence*, why should we suppose that such a fuzzy thing as "common sense" should make things easier on us?

Interestingly enough, several pages after, Spencer-Brown (unwittingly?) resolves the very dilemma that he introduced:

> A major aspect of the *language of mathematics* is the degree of its formality. Although it is true that we are concerned, in mathematics, to provide a shorthand for what is actually said, this is only half the story. What we aim to do, in addition, is to provide a *more general form* in which *the ordinary language of experiences* is seen to rest. As long as we confine ourselves to the subject at hand, without extending our consideration to what it has in common with other subjects, we are not availing ourselves of a truly mathematical mode of presentation. [emphasis added] (ibid., xx)

"[A] more general form [of] the ordinary language of experiences. . . " This is a highly insightful way of interpreting the foundations of mathematics. Such a manner of speaking harkens back to Boole's *Laws of Thought* and such statements as "That *language* is an instrument of *human reason*, and not merely a medium for the expression of thought, is a truth generally admitted." It calls us back to the days of Leibniz's *characteristica universalis* and his ambitions to fashion a language general enough to encapsulate all mathematical, metaphysical and scientific concepts. We can almost hear Gauss reminding us that "algebra is a science of the eye."

This insight of Spencer-Brown is also redolent of Charles Peirce's tightly intertwined notions of *Firstness* and *iconicity*. Peirce's category of Firstness consists of raw qualia (say "yellowness" or "pain"), while his notion of "icon" he describes, rather esoterically, as that *sign* which "partakes of some more or less overt character of its Object." In less esoteric terms, an icon is simply a sign which *resembles* a given object, e.g. the sign "⊖" could be an icon whose object is a pharmaceutical pill.

These two notions dig down to qualitative, perceptual bedrock. One may doubt that he or she is experiencing or perceiving the "real world," but for one to doubt the very existence of this experiencing—what we might call the *givenness* of icons and the Firstness they possess—would no doubt be a grave lapse of "common sense." That is, even if we quibble over whether or not a given set of axioms qualifies as "self-evident," if we cannot even grant the mere *existence* of a subjective *percept*, it seems that, pragmatically speaking, we should have to abandon the practice of mathematics altogether. Should it not be obvious that there could be *no mathematical practice*—no manipulation or generation of abstract properties rendered *perceptually*—without what Spencer-Brown terms "the ordinary language of experiences"?

What I wish to do is this: armed with C. S. Peirce's notions of Firstness and iconicity and the fervent desire to strike logical and mathematical "bedrock," I will see if it is possible to dig *even deeper* than the distinction. This certainly does not mean that I wish to uproot Spencer-Brown's genius insights. Rather, in the spirit of simplicity and elegance, I seek even simpler, more primitive forms of expression that can encapsulate logic and mathematics. To do so, I will also highlight an oft-neglected linguistic system, *Lincos*. Lincos is a rigorous, logically precise language offered in 1960 by the Dutch mathematician Hans Freudenthal, designed to be understood by *any conceivable form of extraterrestrial life.* Although its development and exposition in the original 1960 text can be quite tedious to sift through, there are numerous nuggets of beauty and simplicity to be found in Freudenthal's project, many of which prove useful for fashioning a maximally *self-evident* vehicle of logical expression.

In the next section, "Iconicity in View", I will broach C. S. Peirce's definition of the concept and examine the relationship between his conception of logical iconicity and that of Spencer-Brown. There will be three subsections, "Iconicity in Peirce", "Iconicity in Spencer-Brown" and "Self-Evidence?". In the subsequent section, "Learning How to Learn: Lincos", I fuse the iconic approach to formalism with the heuristic and propaedeutic approaches of Dr. Freudenthal. Finally, in "Seeking Semiotic Self-Evidence", I offer my approach, a synthesis of Spencer-Brown, Peirce and Freudenthal: a new, *self-evident and iconic* vision for logical and mathematical expression.

2 Iconicity in View

In William Bricken's brilliant *Iconic Arithmetic* series, he contrasts *symbolic* mathematics with *postsymbolic* or *iconic* mathematics. He states near the beginning that "[the] objective is to learn more about conventional arithmetic as it unfolds within the unconventional conceptual system of nested containers." Whereas symbolic mathematics is "disconnected from human evolution, from human learning, from human psychology and from our natural human capabilities", *postsymbolic mathematics* brings us back to our roots:

> Very early in our civilization, humans living in city-states put tallies into groups. These groupings eventually became digits. Throughout most of recorded history, counting to determine *how many* was a specialist skill. A cupful of beans did not elicit a desire to know how many beans. Numbers were magical icons; only recently do we count. Symbolic arithmetic focuses our attention on the beans, postsymbolic arithmetic focuses on the cup. (Bricken 2018, xxii–xxiii)

Bricken then goes on to mention the two stars of our show: *George Spencer-Brown* and *Charles Sanders Peirce.* Bricken notes that both fellows "used diagrams and visual thinking extensively," (ibid., 10) and that they "understood that thinking unites imagery with structure." (ibid., 10) It is these shared tendencies of these two intellectual giants that I wish to compare, contrast, and elaborate on.

3 Iconicity in Peirce

What comes to mind when I say the term "mathematical diagram"? Perhaps it is something like this:

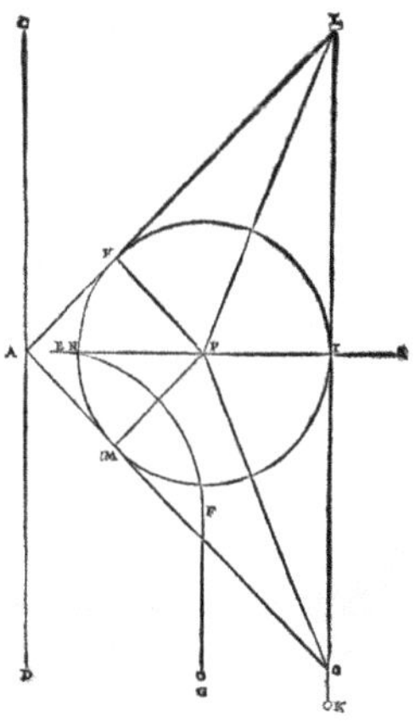

Or, for the more logic-inclined, perhaps it is something like this:

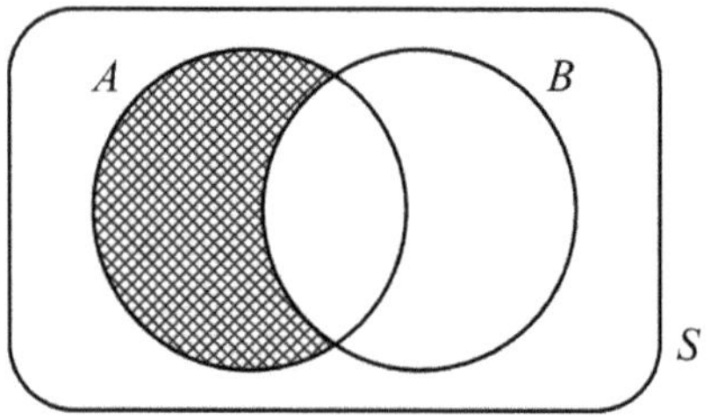

Yes, these are diagrams indeed. But traditionally graphical, pictorial diagrams like these are by no means the only sort of diagrams there are. C. S. Peirce ventures that "[t]he diagrams in which the mathematician's hypotheses are embodied are of two kinds." He then specifies what he means:

> In the one kind the parts of the diagram are seen in the visual image to have the relations supposed. In the other kind of diagrams, the parts have shapes to which conventions or "rules" are attached, by means of which the supposed relations are attributed, or imputed, to the parts of the diagrams. Geometrical figures are diagrams of the inherential kind, while *algebraical formulae are diagrams* of the imputations kind. [emphasis added] (Pierce and Moore 2010, xxix)

"Algebraic formulae are diagrams..."?! This is certainly an interesting, somewhat counterintuitive perspective. Perhaps we can illustrate Peirce's proposition with several examples.

To mathematicians, algebraic symbols are something other than diagrams because they have compartmentalized these little shapes, learning to immediately associate them with something other than a mere shape or squiggle. To *native speakers*, the characters of their learned language have been compartmentalized in much the same way; these characters lose their diagrammatic character simply because they have been acquired as *tools*.

Say you have been presented with the following:

ᐊᓕᐊᓇᐃᒍᓯᒃ

If you speak Inuktitut then unfortunately this example is lost on you! But, if you are one of nearly 8 billion people who do not, then this is a perfect example. For the vast majority of us, this word is just a collection of *shapes* juxtaposed against each other with *no salient associations.* In this sense, we might say that to non-Inuktitut speakers, the above is *not yet a diagram.* It is as if we showed the Venn diagram graphic above to a child who could not yet distinguish square from circle, letter A from letter B. The Inuktitut word above, to the unlearned, is simply an *arbitrary* collection of shapes and squiggles.

We might imagine a language, "Geometrian", the alphabet of which consists of solely polygons and circles, some of which are shaded and some of which are blank. An example of a word in Geometrian could be the following:

This is another example of a pictorial arrangement that is *not yet a diagram.* What then does it take to make a diagram? The answer—in addition to an acquired *meaning*—lies in the *relations* between the shapes and letters involved:

> In order that they may be reasoned about mathematically, [the mathematician's hypotheses which are "always the conception of *a system of relations*"] must be conceived as embodied in some kind of objects; but the character of the objects, apart from the *relations*, is utterly immaterial. *They are always made as bare, skeleton-like, or diagrammatic as possible.* With mathematicians not born blind, they are always visual objects of the simplest kind, such as dots, or lines, or letters, and the like. [emphasis added] (Peirce ca. 1895)

For the sake of a colorful example, assume that we have a *sociologist* instead of a mathematician. Suppose our sociologist is conducting a highly ambitious study on the relationship between paternal parenting styles, masochistic or "self-defeating" personality disorder and different types of religious affiliation (or lack thereof). Say that our sociologist is trying to isolate a certain relationship that exists when a child is raised by an atheist or a-religious father versus a child who was raised by a father who has masochistic personality disorder (and is neither atheist nor a-religious). He or she could use the following Venn diagram to illustrate the relationship:

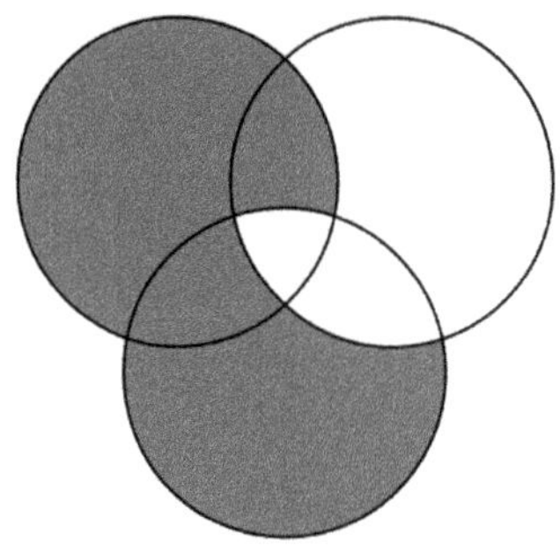

We could have "D" stand for *dad*, "M" stand for *masochistic personality disorder* and "A" stand for *atheistic* or *a-religious*. This Venn diagram possesses certain *relational attributes* useful for our sociologist's investigation: there is relation (1) the collection of all non-masochistic, non-atheistic fathers $\{\boldsymbol{D} \backslash (\boldsymbol{A} \cup \boldsymbol{M})\}$ (a control group thereof), (2) the collection of all non-atheistic people with masochistic personality disorder $\{\boldsymbol{M} \backslash (\boldsymbol{A} \cup \boldsymbol{D})\}$ (again a control group), (3) those fathers who are *not* suffering from masochistic personality disorder but *are* atheistic or a-religious $\{\boldsymbol{D} \cap \boldsymbol{A}\}$ and (4) those fathers who *are* suffering from MPD but are *not* atheistic or a-religious $\{\boldsymbol{D} \cap \boldsymbol{M}\}$.

This Venn diagram thus conforms to the Peircean definition of a diagram, a collection of "visual objects" which "embody" the mathematician's (or scientist's) "conception of a system of relations."

Enough about diagrams then. How exactly does Peirce's notion of diagram square with his notion of *iconicity*? We have only briefly touched on iconicity, simply mentioning that "an icon is simply a sign which *resembles* a given object." Because there has been much hairsplitting over the precise meaning of Peirce's often undulating terminology, I will share a specific, *maximally relevant* excerpt from his "Of Reasoning in General" (1895, in Peirce 1998):

> An *icon* is a sign which stands for its object because as a thing perceived it excites an idea naturally allied to the idea that object would excite. Most icons, if not all, are *likenesses* of their objects [...] A diagram is a kind of icon particularly useful, because it suppresses a quantity of details, and so allows the mind more easily to think of the important features [...] Many diagrams resemble their objects not at all in looks; it is only with respect to the relations of their parts that their likeness consists. (Peirce 1998, 13)

Peirce tells us that a diagram is "a kind of icon," one which is especially informative due to its skeletal, bare-bones nature that "suppresses a quantity of details," allowing one to "more easily think of the important features." This reminds us of another of his assertions, namely, that "[The mathematician's hypotheses which are 'always the conception of *a system of relations*'] must be conceived as embodied in some kinds of objects... They are always made as bare, skeleton-like, or diagrammatic as possible." (Peirce ca. 1895)

But what about his mystical-sounding assertion that an icon "stands for its object because as a thing perceived it excites an idea naturally allied to the idea that object would excite?" Well, just as we often use words to explain other words, we might as well use an *icon* to better explain Peirce's notion of *icon*! Consider the following:

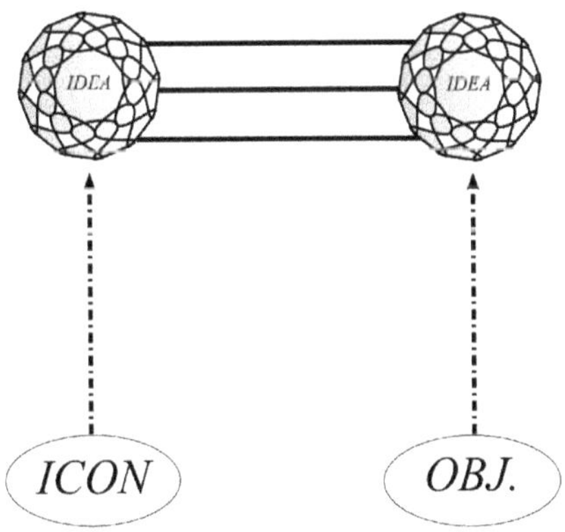

Although Peirce's verbiage makes an icon sound quite obscure, the idea is very simple: when one encounters an *icon*, ideas arise in one's mind that are similar to those ideas experienced while encountering the *object itself.* There is some sort of deep, structural similarity between the *associated ideas* of the *icon* and those of the *object.*

Peirce used this precise notion as he forayed into the graphical representation of logic. Peirce's "existential graphs," though long denigrated and only "unearthed" to the philosophical community many decades after his death, represent an ambitious attempt to *iconicize* logic. Although there is no hope of covering his approach in great detail, I will broach the basics of his approach—in addition to some potential pitfalls—before we discuss George Spencer-Brown's approach to iconicity.

Whereas Spencer-Brown urges us to "draw a distinction" through a background space, C. S. Peirce talks of "scribing a graph" on a "sheet of assertion" (Roberts 1973, 31). Peirce urges the reader to consider the sheet of assertion (whatever surface one is marking on) as a sort of dynamically evolving *universe of discourse.* Whatever we decide to write upon our sheet of assertion, we are making an addition to the universe or domain in front of us. Rather obscurely, Peirce would claim, "If the [sheet of assertion] be blank, this blank, whose existence consists in the absence of any scribed graph, is itself a graph" (ibid., 32).

Peirce built his graphs for propositional logic on the two connectives, $\wedge$ and $\neg$ (which, as a pair, are functionally complete). He only employed one sort of icon (or, more accurately, *sub-icon*)—a circle or a "cut." (Roberts 1973, 28) To draw a "cut" around a proposition was to *negate* it. To *aggregate* different propositions (to assert $X \wedge Y$ for any two propositions) Peirce simply relied on spatial proximity; if two propositional letters X and Y were next to each other, with no circles drawn around them, one would be justified in asserting the truth of "$X \wedge Y$." Take for example the following:

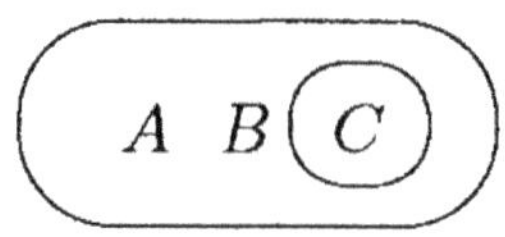

This graph would amount to the sentence "It is false that A and B and not-C" or $\neg(A \wedge B \wedge \neg C)$. Since there is a cut drawn around C, it is negated. And, since there is a cut drawn around both A, B and $\neg C$, these portions are also negated. Another example would be:

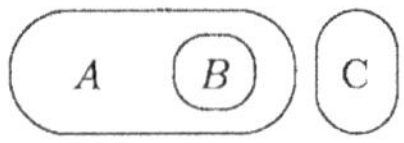

This graph amounts to (at least) each of the following:

- "If A then B, but not C"
  - "It is not the case that A and not B, and it is also not the case that C"

and

- "$[(A \rightarrow B) \wedge \neg C]$"

Peirce was quite proud of these graphs, convinced that this new diagrammatic approach to logic was not only novel but that it would come to reveal *unexpected logical truths.* He noted that "a great distinguishing property of the icon is that by the direct observation of it other truths concerning its object can be discovered than those which suffice to determine its construction." He also went on to compare such graphs to maps, hopeful that such logical maps would help future logicians and mathematicians explore uncharted terrain.

To tie these graphs back into Peirce's notion of the icon, we can offer a "new and improved" form of the visual aid we used earlier:

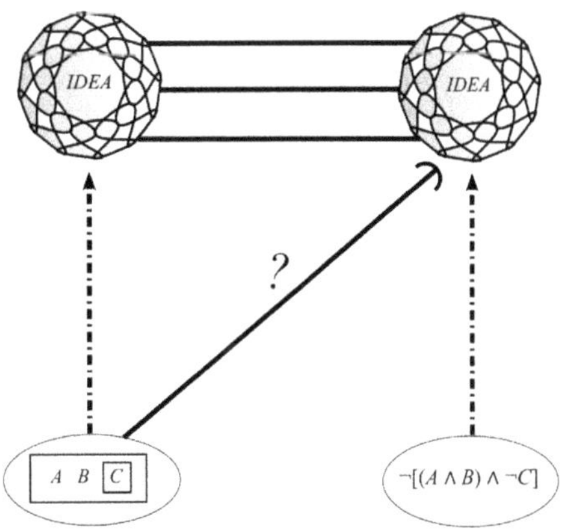

This is the same basic model of Peircean iconicity as before, with slight changes. Before we had "Icon" in the bottom left oval and "Object" in the bottom right. Now we have a Peircean existential graph as *icon* and a conventional, symbolic logic formula as *object.* This graphic represents the general idea of what Peirce had in mind with his existential graphs. He tried to fashion a maximally iconic form of logical representation, that which would elicit *the very same ideas* as do traditional, symbolic formulas. A logician looking at a Peircean graph should, ideally, shout "Aha! I recognize this form!" and promptly be able to produce its symbolic counterpart. In this sense, the conceptual "gap" between the bottom left oval and the bottom right should be minimal. But there is one issue. The diagonal line running from the bottom left oval to the right-hand idea is labeled with a *question mark* for a very good reason.

As I mentioned earlier, the whole point of iconic representation is to evoke in the reader those *very same ideas* which are evoked by the traditional formalisms. When one thinks up a "recipe" for a certain icon, one must *not only* consider one's *subjective* ideas associated with the object in question; one must also consider what *relational qualities* the idea should have generally, or *objectively*, speaking. This is an inherent problem in any society in which genuine "mindreading" has not yet hit the technological stage! As Charles Peirce was not privy to the subjective contents of other logicians' minds, he could only make *educated guesses* as to how he should design his existential graphs. In this sense, the creation of such icons is an *empirical*, as opposed to a *rational*, game.

Interestingly enough, Peirce tangentially touches on such issues. But he does not seem to have fully taken these difficulties into account. As Roberts notes in his *The Existential Graphs of Charles S. Peirce*, he asserted that his existential graphs enabled one to reproduce the "very motions of reasoning," "the operation of thinking *in actu*." Yet, Roberts sharply notes that by "motions of reasoning" Peirce did not mean the thought-process that "takes place in the mind." He quotes Peirce as saying:

> A man goes through a process of thought. *Who shall say what the nature of that process was? He cannot*; for during the process he was occupied with the object about which he was thinking, not with himself nor with his motions. [emphasis added] (Roberts 1973, 113)

Roberts notes that Peirce held that "[what] is relevant to logic is the argument formulated after the fact as a summing up of the thinking-process," and that logical propositions and arguments "constitute a kind of 'self-defence' of the original process." (ibid., 113)

There we have it... The fundamental contradiction, so to speak. Peirce himself admits that the actual process of reasoning is quite tricky, if not impossible, to grasp. Yet, in adhering to his definition of an icon, he put himself in a difficult position: quantifying over these varied, dynamic and shifting ideas associated with his objects in question (traditionally formalized logic). In the section "Seeking Semiotic Self-Evidence" I circle back to this shortcoming of Peirce's approach and offer some exciting potential ways around the impasse. This is, however, not meant to lambaste his inventive work. His steps in the direction of a fully iconic logic are extremely admirable.

Having touched on Peirce's notion of iconicity and the mechanics of his iconic logic, we can now move to George Spencer-Brown's project, not unlike Peirce's in both profundity and scope.

4 Iconicity in Spencer-Brown

Before we dive right into Spencer-Brown, I should say a bit about my intent. Spencer-Brown expended a great deal of effort on developing an inferential, algebraic *transformation system.* He was not solely focused on semiotic representation and iconicity; he was also focused on *the deductive functionality of his systems.* In my investigation I am not concerned with the rules of transformation whatsoever; I am concerned solely with striking *semiotic, representationally self-evident* bedrock. For this reason I will be focused solely on Spencer-Brown's *means of representation*, as opposed to his wide-ranging and inventive set of transformation rules.

William Bricken, in the first volume of his *Iconic Arithmetic,* contends that the "academic analysis, enthusiasm, controversy and rejection of Spencer-Brown's work is widely based on a severe misunderstanding that *Laws of Form* describes conventional logic, which it does not." He then wryly notes that the work "becomes much more controversial when it is taken for what it actually is: a postsymbolic foundation for rigorous thinking."

I am inclined to agree with Bricken's assertion, and I wish to briefly investigate Spencer-Brown's program from the perspective of iconic logic. In contrast to the previous section, we will jump right into the specifics of the applications to logic. C.S. Peirce had much to say on the nature of diagrams and icons themselves, while Spencer-Brown—although he offers a beautifully detailed introduction—does not discuss the more philosophical, semiotic aspects of such things at length. What I will do is utilize the graphics from the previous section, the "icon of an icon" to illustrate what Spencer-Brown's approach does and does not achieve.

On page 91 of the 2011 edition of *Laws of Form*, Spencer-Brown offers the following notation for propositional logic:

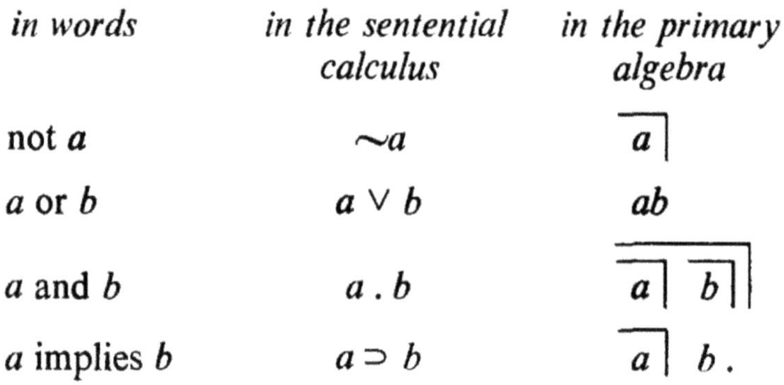

in words	*in the sentential calculus*	*in the primary algebra*
not *a*	~*a*	*a*⌉
a or *b*	*a* ∨ *b*	*ab*
a and *b*	*a* . *b*	*a*⌉ *b*⌉⌉
a implies *b*	*a* ⊃ *b*	*a*⌉ *b* .

In order to show the similarities of Spencer-Brown's notation with that of Peirce, we might as well render the "cross" as a fully enclosed *box*, in the vein of Peirce's so-called *cuts*:

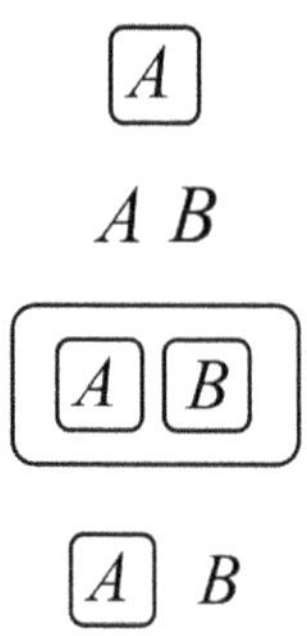

For both Spencer-Brown and Peirce, "(A)" represents "$\neg A$." However, this is where the similarities end. While "AB" represents *disjunction* for Spencer-Brown, for Peirce it represents *conjunction*. While "((A)(B))" represents *conjunction* for Spencer-Brown, it represents *disjunction* for Peirce (the negation of "($\neg A \wedge \neg B$)"). And, whereas "(A)B" is conditional implication for Spencer-Brown, it is "$\neg A \wedge B$" for Peirce.

Like I mentioned several paragraphs ago, I do not wish to focus on the rules of transformation of Spencer-Brown's or Peirce's system.

Rather, the question I have in mind is this: say we printed out the above graphic and handed it to a Freshman "Intro to Logic" student... would he or she, *without prior knowledge* of either Spencer-Brown or C. S. Peirce, immediately—through some supposed "self-evidence" of the diagrams—affix an interpretation to each diagram coinciding with either Peirce's or Spencer-Brown's interpretations of them (or provide a privileged interpretation and reasoning therefor)? I strongly suspect the answer is, "No way!" What this tells us is that, despite the profound inventiveness of both thinkers' diagrams, genuine representational *self-evidence* has yet to be reached.

Spencer-Brown adheres to the same basic project that Peirce does: he is using iconic, diagrammatic representation, attempting to "excite" the very same ideas in the mind of the reader as those excited by traditional, symbolically rendered logic and algebra. In this sense, we can offer an improved graphic:

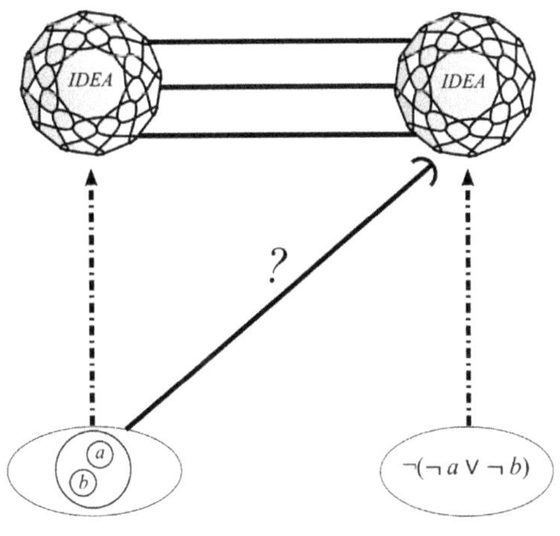

Let me reintroduce something I mentioned in conjunction with the last diagram of this kind:

> The diagonal line running from the bottom left oval to the right-hand idea is labeled with a *question mark* for a very good reason [...] When one thinks up a "recipe" for a certain icon, one must *not only* consider one's *subjective* ideas associated with the object in question; one must also consider what *relational qualities* the idea should have generally, or *objectively*, speaking [...] In this sense, the creation of such icons is an *empirical*, as opposed to a *rational*, game.

While it is the case that Spencer-Brown's mark and distinction represent a brilliant departure from traditional, symbolic notation, his approach falls victim to the same barrier of self-evidence as did Peirce's existential graphs—in crafting an icon based on logical or mathematical symbolism, one cannot know *a priori* exactly which ideas (in a generalized, *subjective* sense) one is trying to evoke. One may certainly make an educated guess, a guess which works marvelously for all *practical* intents and purposes, but this does not entail that the icons in question are indeed self-evident representatives of their objects. Iconic, idea-based self-evidence is largely a matter of *intersubjectivity*.

5 Self-Evidence?

For all this talk about "self-evidence," I have admittedly said very little about *what* exactly *it is*. I said in the opening section that I seek "even simpler, more primitive forms of expression that can encapsulate logic and mathematics." But in what sense is this to be taken?

From the outset we can delineate two main types of iconic self-evidence (though these need not be the only kinds), what I will call *operational self-evidence* or *OSE* for short and *representational self-evidence (RSE)*. We can illustrate the first kind of self-evidence by considering Spencer-Brown's so-called primitive equations of his "calculus of indications":

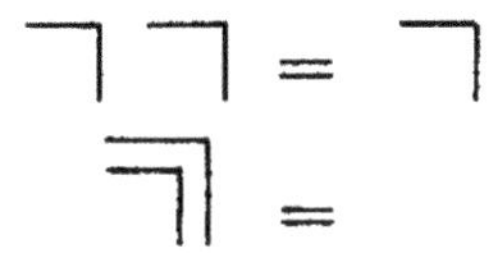

These equations *by themselves* are *not* self-evident. When these two rules are given *without context*, it could be safely argued that there is no self-evidence to be found at all. What I mean by "operational self-evidence" takes into account the *surrounding context* of the text in question. Say that, after learning the above equations, we were given the following problem to solve:

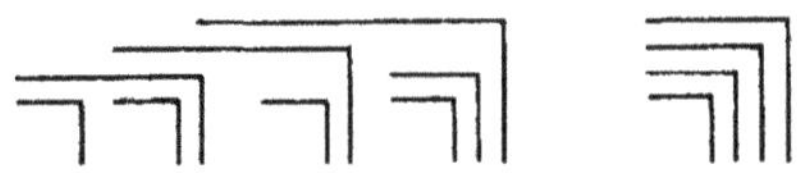

Having already committed the two "primitive equations" to memory, we would be able to *self-evidently* perform the following sequence of reductions:

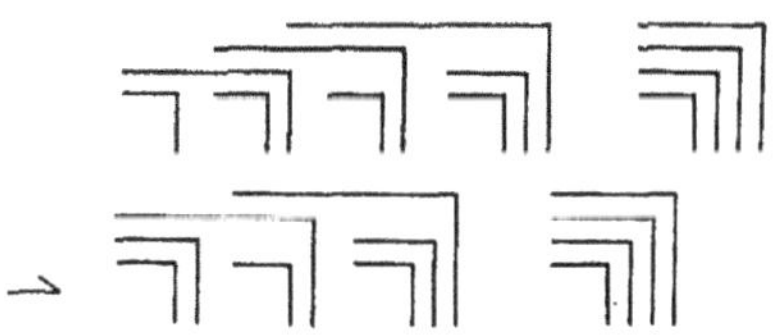

This is *operational self-evidence* in action. Where we take the Peircean route and consider the pages of Spencer-Brown's text to be sort of "sheet of assertion," an already-composed universe of discourse, we reach the following general form of operational self-evidence:

Here "S.A." stands for *sheet of assertion*, the black shapes are *symbols* (usually *words of some natural language*) and the *icons* in question occupy the *symbolic context*. With this general picture in mind, we can offer a precise definition of operational self-evidence:

> ***OSE*** *Call a given group of icons and their accompanying rules of transformation operationally self-evident when—given the symbolic context—one is presented with a well-formed arrangement of these icons, it is immediate from the visuospatial form alone how to apply each of the applicable rules of transformation.*

There is no doubt that Spencer-Brown's calculus of indications adheres to the criteria of *OSE*. It is in this sense that we can call the mark and its manipulations "self-evident." However, this is only one half of the "self-evident story." To utilize C. S. Peirce's comparison of his existential graphs to geographical maps, we might say that both Spencer-Brown's and Peirce's iconic notations are *maps* which require their respective *keys*. Without these keys, there is little self-evidence to be found.

What I mean when I say that I am seeking a "maximally *self-evident* vehicle of logical expression," is in line with the other variety of self-evidence: *representational self-evidence* or *RSE*.

To use the example with Peirce's "sheet of assertion" again, we can assert that representational self-evidence has the following general form:

S.A.

Icons

If an icon or a grouping thereof is representationally self-evident, then there is *no symbolic context required whatsoever.* The only thing required in such a case is the viewer's acquaintance with the represented ideas in question. In the case of propositional logic, if we had fashioned a grouping of icons to represent the sentence "$A \wedge B$" all that would be required for *RSE* is a conceptual idea corresponding to "$A \wedge B$" and the ability to perceive the icons upon the sheet of assertion. This is the sort of self-evidence I have in mind:

RSE *Call a given group of* icons *representationally self-evident if, when one is presented with a well-formed arrangement of these icons, it is immediate from the visuospatial form alone—without any symbolic context necessary—which idea(s) are intended to be represented.*

Given this definition, we can assert the following: neither C. S. Peirce's nor George Spencer-Brown's version of iconic representation—though they are both operationally self-evident—is *representationally self-evident.* It is representational self-evidence that I wish to achieve.

But how on earth can we achieve such a thing? How is it possible that some grouping of icons can stand alone, without any symbolic context whatsoever, and still be understood to represent such complex ideas as those found in logic and mathematics? To help see how this feat can be accomplished, we can look towards a very exciting and rather recent phenomenon—the creation of so-called *extraterrestrial languages* and the inchoate field of *Astrolinguistics.*

6 Learning How to Learn: Lincos

Shall we start with yet another diagram? Many folks have likely seen this:

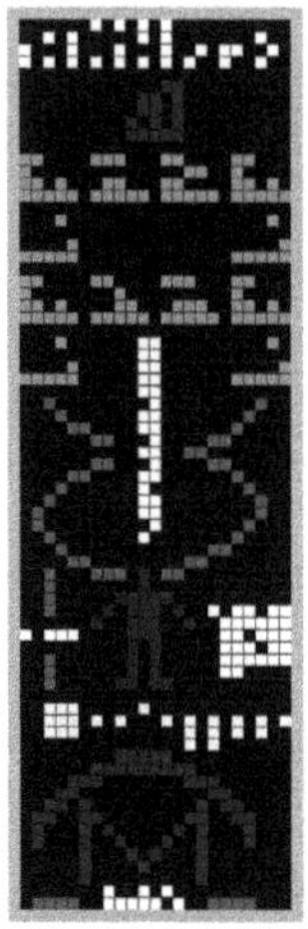

It is somewhat common knowledge that this graphic was a *message*, a message for an *extraterrestrial audience*, beamed from the Arecibo Telescope in Puerto Rico in 1974. At least, that is how the media liked to interpret it! In reality, the graphic was only meant to be a flexing of mankind's technological muscles; it was not a genuine "message" at all. The Cornell professor of astronomy at the time, Donald Campbell, stated "It was strictly a symbolic event, to show that we could do it..."

As the article, "It's the 25TH Anniversary of Earth's First (And Only) Attempt to Phone E.T." says, "The real purpose of the message was to call attention to the tremendous power of the radar transmitter newly installed at Arecibo and the ability of the telescope's 1,000-foot diameter dish antenna to project a powerful signal into space."

All the same, the "Arecibo message," as it is called, is an *icon*. The pixelated images adhere to C. S. Peirce's notion of iconic form. Each of the concepts addressed in the Arecibo message—from numerals to the structure of DNA and a representation of the Arecibo telescope itself—attempts to "excite" *those very same ideas* which are excited by those objects (whether mental or physical) themselves.

Another well-known example of proposed extraterrestrial communication is the "Pioneer plaque":

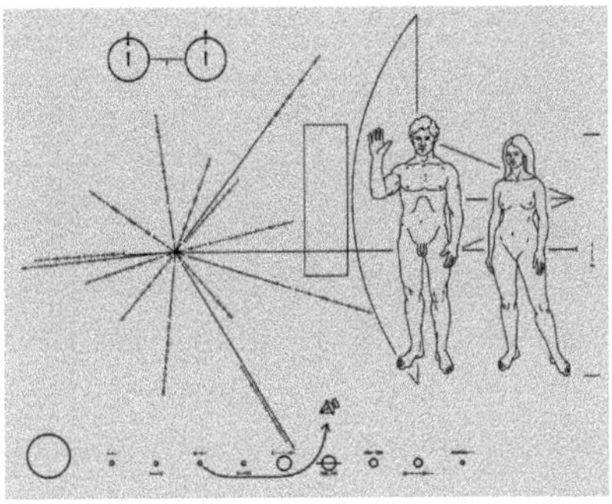

Two of these plaques were made. One was placed aboard the 1972 Pioneer 10 spacecraft and the other aboard the 1973 Pioneer 11. Just like the Arecibo message, each of the concepts on the plaque is represented by an *icon*. And this is *no accident*.

Reflecting back on our discussion of Peircean "sheets of assertion" and the criterion of "representational self-evidence" will help us gain a clearer picture.

7 Extraterrestrial Representation

In the previous section I made the claim that both Charles Peirce's and George Spencer-Brown's work on iconic representation followed the following general form:

This is the exact same "general form" that we find in any traditional academic treatise involving iconic representation (ironically enough, including this paper!) or any generic extended description and introduction of iconic representation. The black shapes above and beneath "Icons" represent words of natural language and other generic symbols. In the previous section, I termed this the "symbolic context."

However, when we consider an audience which is *not acquainted with any known natural language* or human symbols (like supposed extraterrestrial beings), there is no use for symbolic context! Imagine if the creators of the Pioneer plaques decided to use English or Ukrainian or some other *human natural language* to communicate. To do so would be an arbitrary and self-defeating task. Rather, if we want to communicate with an audience without any symbolic context whatsoever, we are left with *icons* (and their constituent markings or components) the whole way down:

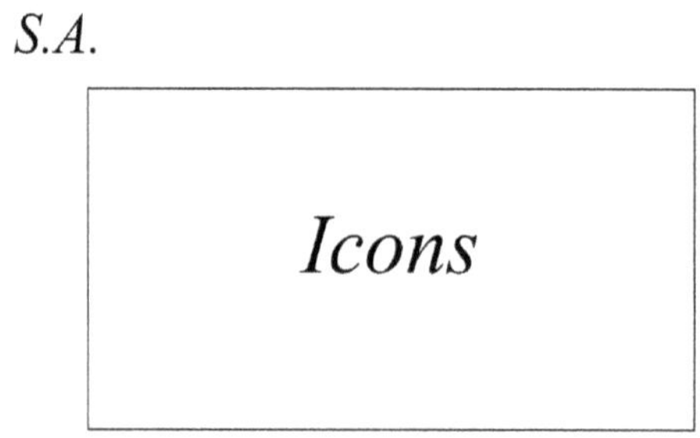

This basic notion of iconic communication absent any symbolic context is the cornerstone of the burgeoning field known as *Astrolinguistics*.

A full decade before both the Arecibo message and the Pioneer plaques, the first book-length exposition in the field of Astrolinguistics was published by the Dutch mathematician Hans Freudenthal. *Lincos: Design of a Language for Cosmic Intercourse* is a 224-page, tremendously dense feat of logico-mathematical prowess and creativity. In the book's lengthy introduction, Freudenthal discusses the importance of iconic expression and semiotic self-containment:

> Lincos has to be taught to the receiver. Therefore our program to be broadcast cannot have the character of a newsreel. On the contrary, in the beginning we shall communicate facts which may be supposed to be

> known to the receiver. In this way we can try to facilitate the work of deciphering.
>
> As our means of showing are heavily restricted, we cannot start with a concrete subject matter. Mathematics is the most abstract subject we know ... So we have decided to start our program with mathematics. (Freudenthal 1960, 21)

In other words, symbolic context such as natural language will be of no use. The very system of reference and communication must be built from scratch in a self-contained fashion. To recall some of the previous terminology I used, we might say that the aim of Freudenthal's project was to achieve *representational self-evidence* as so defined:

> ***RSE*** *Call a given group of* icons *representationally self-evident if, when one is presented with a well-formed arrangement of these icons, it is immediate from the visuospatial form alone—without any symbolic context necessary—which idea(s) are intended to be represented.*

Yet, to say that Freudenthal's *only goal* was to achieve RSE as defined above would be misleading. Remember the following assertion of Freudenthal: *Lincos has to be taught to the receiver.* Also notice the following phrase from the definition of RSE: *a well-formed arrangement of icons.* The definition I offered for representational self-evidence glosses over the issues of *teaching* and *learning* entirely; it assumes that either (1) the reader is already familiar with what makes the given arrangement of icons "well-formed" or (2) that there is no learning or teaching process at all, and that the iconic representation takes place in some sort of social and communicative vacuum.

For these reasons, we can offer an even more precise definition of the breed of self-evidence that Dr. Freudenthal strove to deliver with Lincos. I will call this kind of self-evidence *generative representational self-evidence* or *G-RSE*:

G-RSE *Call a given group of icons generatively representationally self-evident if—after learning what constitutes a "well-formed" grouping of icons without any assistance from symbolic context—when one is presented with a well-formed arrangement of these icons, it is immediate from the visuospatial form alone—without any symbolic context necessary—which idea(s) are intended to be represented.*

I use the term "generative" to impart the dynamic and progressive building-based nature of projects like Lincos.

8 Extraterrestrial Communication

On the theme of learning and communication-based iconic form, we can return to something I have emphasized several times throughout this paper:

> [I]n crafting an icon based on logical or mathematical symbolism, one cannot know *a priori* exactly which ideas (in a generalized, *subjective* sense) one is trying to evoke. One may certainly make an educated guess, a guess which works marvelously for all *practical* intents and purposes, but this does not entail that the icons in question are indeed self-evident representatives of their objects. Iconic, idea-based self-evidence is largely a matter of *intersubjectivity.*

Let me repeat: *iconic, idea-based self-evidence is largely a matter of intersubjectivity.* Let us briefly revisit *why* this is the case.

In the section, "Iconicity in Peirce" I offered the following graphic as a representation of what a Peircean icon is:

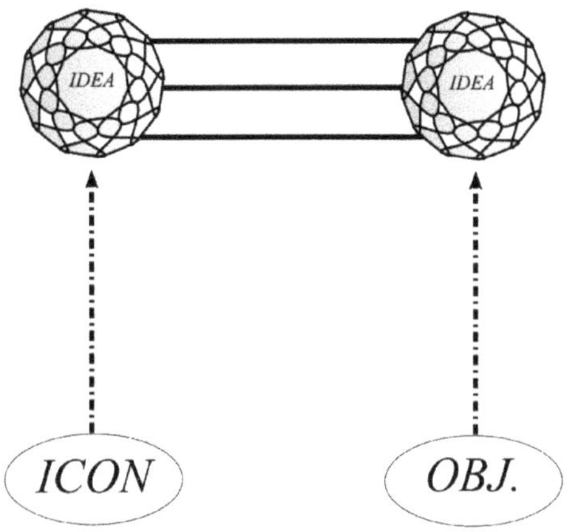

This captures Peirce's assertion that "an *icon* is a sign which stands for its object because as a thing perceived it excites an idea naturally allied to the idea that object would excite."

Both Peirce and Spencer-Brown, in their respective projects, set their sights on *conventional, symbolic logic formulas* such as "$\neg[(A \wedge B) \wedge \neg C]$ as the *objects* which they intend to *iconically represent*. For this reason, we represented C. S. Peirce's approach, his existential graphs, in the following way:

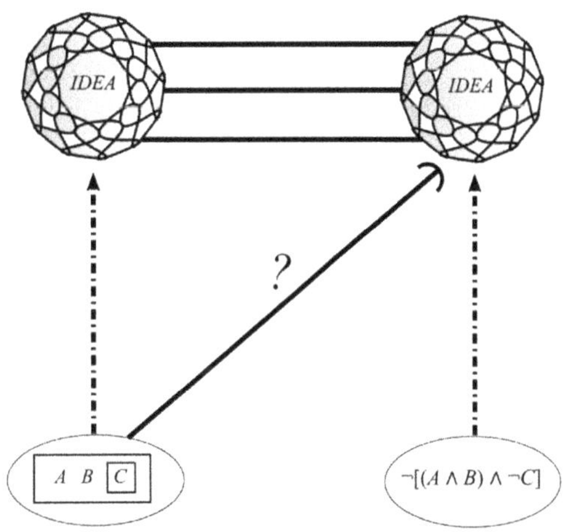

The diagonal line labeled with the question mark represents the attempt at iconic representation of an *idea* allied with a specific formula of traditional symbolic logic. This is a tremendously difficult and messy process, especially for an object as abstract as a logico-mathematical formula. As I mentioned previously, "One may certainly make an educated guess [as to the proper iconic representation], a guess which works marvelously for all *practical* intents and purposes, but this does not entail that the icons in question are indeed self-evident representatives of their objects."

The downside to both Peirce's and Spencer-Brown's approach to iconic representation is that there is no real-time *communication* between the author and the reader, the "sender" and the "receiver." If we admit that the process of striking upon a self-evident iconic representation for logic and math is fundamentally *intersubjective* and *empirical*, then we must also admit that real-time, back-and-forth communication is absolutely necessary. This is what Hans Freudenthal recognizes and addresses in his masterful *Lincos*.

9 Learning Lincos

Whereas both C. S. Peirce and George Spencer-Brown used natural language to introduce, explicate and advance their breeds of iconic representation, if we were to take an "extraterrestrial's eye view" of Hans Freudenthal's Lincos, we would see only a succession of *objects* and *signs*. Of course, assuming no prior acquaintance with any given sign, the signs themselves would have to be *built, generated* from assorted *objects*.

We can use a series of graphics to illustrate the basic process of the Lincos style of communication.

Firstly, one starts with a generic object:

$$OBJ_i$$

Literally any sort of intricate design, shape, marking or meandering squiggle will do. Curiously enough, in Freudenthal's case, he introduces his *signs* at the outset *as objects.*

For example (from Freudenthal, 1960, 45):

> **1 01 0.** # > | < | = | + | − | ≠ | ≦ | ≧ |
> · | · · | · · · | · · · · · · · · |
> 1 | 10 | 11 | 111 |
> *a* | *b* | *c* |
> → | ∨ | ∧ | ? | ↔ #   ·
>
> **Loose Lincos words are presented, without any context, in order to stress their individuality. So it will be somewhat easier for the receiver to recognize them when they occur in a certain context.**
> **The bold-faced strokes mean pauses.**

Freudenthal introduces the mathematical signs ">", "<", "=", "+", "−", "≠", "≦" and "≧" at the outset, separated by *pauses* represented by "|". Freudenthal takes the liberty to call these "words." Yet, in this stage of the communication, each of the usual mathematical signs he lists are mere *objects.* These markings have yet to acquire *distinctive and stable meaning* in the mind of the interlocutor. We can impart this intuition by showing an illustration of two generic objects (and the resulting composite object) which have yet to be agglomerated into a sign:

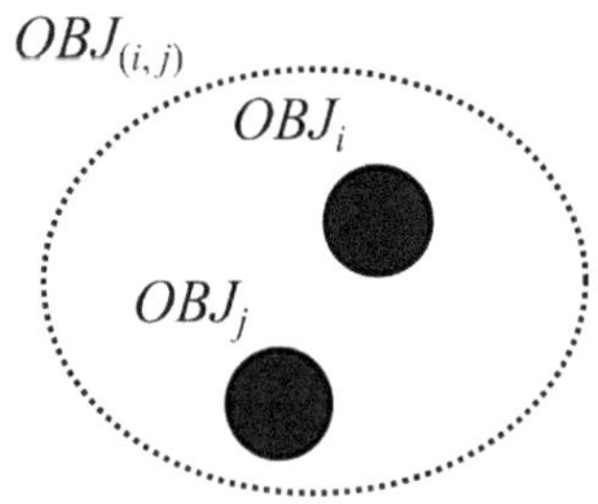

This much is quite simple. It is intuitive that one must start with objects if one wants to build icons from scratch. But how exactly do we make the leap from mere object(s) to *bona fide* signs? Freudenthal tackles this question straight away (Freudenthal 1960, 45):

> **1 00 0. Pairs of signs * will enclose the printed image of a program text. A (metatextual) "and so on" after a text indicates that this text is an exemplary extract from the factual program. When carrying out the program we will replace this text by a large number of texts similar to the text we have printed. If the number of examples is large enough, we may expect that the receiver can generalize the quasi-general definition or proposition that is intended by the program text.**

Freudenthal's answer to the question is a refreshingly pragmatic one: *repetition, repetition* and then *more repetition*!

Before this excerpt, in his nearly 50-page introduction, he offered the following congruent line of thought:

> We shall communicate a number [of formulas] that is large enough to justify the supposition that an intelligent receiver can conclude from these texts what the words (complexes of signs) +, =, >, < mean [...] Children learn arithmetics by a finite number of instances, and a set of rather unformal rules. They succeed in generalizing the instances in a unique way. (Freudenthal 1960, 21)

Simply put: the gap between object and icon is to be closed via brute force *repetition* and semiotic exposure. Freudenthal stresses numerous times that this is how children learn not only natural language but mathematics as well. Aggregating from objects to signs through examples is a universal human process of learning.

Alas, it would be an oversimplification to say that all we need to generate icons is the repetitive assemblage of objects. In fact, between objects and icons, there is an intermediate level: that of *indices*. Let us look at Peirce's definition of the term:

> [If the sign in question is] determined by continuity of space and time, it only can be conveyed by first making it (if it be not already) an

> experience common to deliverer and interpreter, and second by forcibly directing the attention of the interpreter to it. A sign which fulfils this function, like the pointing of a finger, or an exclamation like "hi!", "see!", or a pronoun like "I", "you", "this", or a proper name, as yard, kilogram, or a word relative to the position of the interpreter, as "on high", "clockwise", "yesterday", I have called an *index.* (emphasis added) (Peirce and Moore 2010, 52)

We can see an example of such an "index" emerging in the Lincos communication of quantity and equality (Freudenthal 1960, 46):

> **1 02** 1. * · = 1
> · · = 10
> · · · = 11
> · · · · = 100
> · · · · · = 101
> · · · · · · = 110
> · · · · · · · · · · · · · = 1101 *
>
> and so on.
>
> **Ostensive numerals are superseded by *algorithmic* ones, composed of *syllables* written 0 and 1 according to the rules of the *dyadic positional system*. For the convenience of the terrestrial reader we shall sometimes use the decimal code, but as a matter of fact such occurrences should be translated into the dyadic code.**

On the left-hand side we see dots as objects, and on the right-hand side we see numerals. But what exactly is the nature of the familiar equals sign? Is it an *object*? An *index*? An *icon* perhaps? The answer to these questions is not quite as straightforward as we would like to expect! Rather cryptically, we might respond to each question: *it evolves.* The "=" goes from *object* to *index* to constituent of *icon.*

Let us re-examine something I already mentioned:

> [If] we were to take an "extraterrestrial's eye view" of Hans Freudenthal's Lincos, we would see only a succession of *objects* and *signs.* Of

course, assuming no prior acquaintance with any given sign, the signs themselves would have to be *built, generated* from assorted *objects.*

If we are to put ourselves in the shoes (or, rather, *psyche*!) of an extraterrestrial being, we realize that—for the first several examples at least—what appears to us as the familiar "=" sign can be nothing other than a mere *object.* To better see how this is the case, we can revisit an example from our "Iconicity in Peirce" section:

ᐊᓕᐊᓇᐃᒍᓱᒃ

Although this is a word of the Inuktitut language, to a vast majority of humans, this "word" is simply an arrangement of assorted *objects* or *images.* Suppose that, instead of having "· = 1" as the first example, Freudenthal had used the following isomorphic arrangement of objects:

ᐊᓕᐊ   ᓇ   ᒍᓱᒃ

Again, although each of these objects is a member of the Inuktitut alphabet, to most humans the aggregate is simply a jumble of assorted shapes and squiggles. To an extraterrestrial mind yet to be acquainted with the sequence of objects "· = 1", it would appear in much the same manner as Inuktitut does above.

Having established the initial nature of "=" we can now ask: so how exactly does this object become a sign? More specifically: how does "=" make the journey from *object* to *index*? And how does the *icon* arise from the index?

Interestingly enough, it *cannot* be determined *a priori* when "=" transitions from object to sign. Such transition depends on the mind of the extraterrestrial agent in question and its acquisition of meaning. What we can say, however, is that a given object transitions from object to *index* whenever the observer recognizes that some sort of salient or meaningful *connection* can be discerned in the arrangement(s) of objects he or she is presented with. As the observer is trying to discern the precise nature of this "connection", the pattern remains a collection of *objects* and *indices*; *icons* are yet to be found.

Finally, once the observer cracks the code, he or she is able to see given arrangements of objects and indices as constituents of particular *icons.* More specifically, the equals-sign-shaped object in Lincos's lexicon becomes a constituent of an icon once the extraterrestrial(s) in question has, as Freudenthal himself puts it, "succeed[ed] in generalizing the instances [which include =] in a unique way."

We can illustrate this evolution from object to index to component of icon like so:

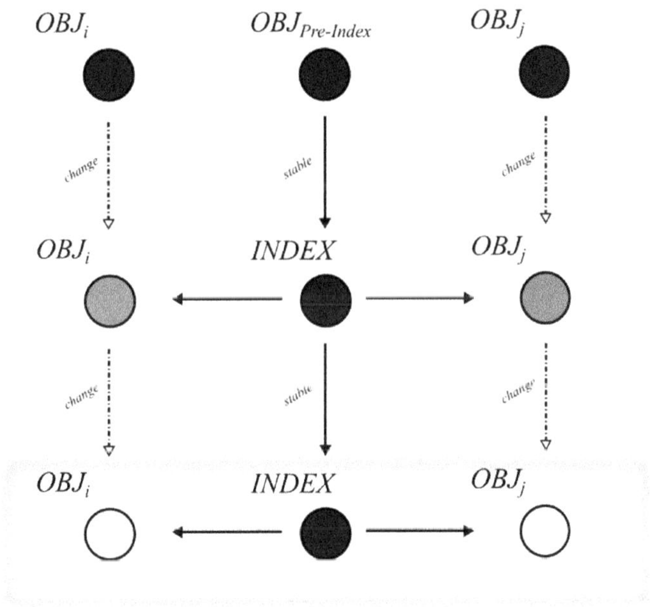

Here we can see an arrangement of three different objects. This diagram is simply an abstract rendering of the Lincos communication starting with the sequence "· = 1". Moving down on the diagram, the object on the left and right side can become *composite objects* (as they do in our Lincos example above, e.g. from "· = 1" to "·· = 10"). The change of color simply signals this possibility of change or composition. The faded gray box at the bottom represents the fact that both objects and the given index have semiotically evolved into an *icon.*

During the "Pre-Index" stage of the object in question (the object "=" in our Lincos example) has yet to be aggregated into an image with any salient or meaningful connections. To the extraterrestrial, the sequence of objects "· = 1" is, at first blush, a seemingly random amalgamation of visual stimuli.

Then, during the "Index" stage of the object "=" it becomes apparent to our extraterrestrial being that the sequence "· = 1" is not merely a random amalgamation. To the extraterrestrial, the equals sign object appears to perform a *function.* There is a pattern to be found in the connection between the two sides of the "=" object. Alas, it is not in this stage apparent to our E.T. friend what exactly this pattern is and how to generalize it.

Between the "Index" stage and the final stage in which the index is a constituent of an icon, we might also consider an immediate stage in which our E.T. has discovered aspects of the pattern but has yet to discover how to fully *generalize* this pattern. A young child who can tell you that $2 + 2 = 4$ and $4 + 4 = 8$ yet cannot yet tell you that $8 \times 2 = 16$ is an example of someone at this stage of the semiotic learning process.

Then, in the bottom stage with the faded gray box, our extraterrestrial has succeeded in fully *generalizing* the pattern with the template "{Dot(s)} = {Numeral(s)}".

Although Freudenthal continues with nearly 200 more pages of examples of "program texts" like the one with the template "{Dot(s)} = {Numeral(s)}", they all follow the same basic process of semiotic evolution presented in the graphic above. Every one of Freudenthal's Lincos program texts is built upon the notion of *generatively representational self-evidence*:

> ***G-RSE*** *Call a given group of icons generatively representationally self-evident if—after learning what constitutes a "well-formed" grouping of icons without any assistance from symbolic context—when one is presented with a well-formed arrangement of these icons, it is immediate from the visuospatial form alone—without any symbolic context necessary—which idea(s) are intended to be represented.*

Yet, despite this astounding display of semiotic self-containment and self-evidence, one cannot help but wonder—especially if one has been exposed to C.S. Peirce's existential graphs and George Spencer-Brown's distinction—if the program texts of Freudenthal's Lincos are as self-evident as they could be. As advanced and streamlined as Lincos is, could there not be a way to push the boundaries of self-evidence even further?

10 Pushing the Boundaries?

Back in the subsection "Self-Evidence?", I mentioned that I was seeking to achieve something I called representational self-evidence:

> ***RSE*** *Call a given group of* icons *representationally self-evident if, when one is presented with a well-formed arrangement of these icons, it is immediate from the visuospatial form alone—without any symbolic context necessary—which idea(s) are intended to be represented.*

When I stated that I was seeking to achieve *representational self-evidence*, this was a bit of a simplification. What I would like to achieve is something closer to *RSE in addition to G-RSE*; I am seeking a fusion of both.

We can now get into some subtleties involving the relationship between representational self-evidence (*RSE*) as defined above and Hans Freudenthal's style of self-evidence, what I have termed generatively representational self-evidence (*G-RSE*).

Firstly, I would like to follow Hans Freudenthal's Lincos in attempting to establish *G-RSE*, teaching the observer(s) what constitutes a well-formed icon and subsequently having this well-formed icon be self-evident. But I would also like to ensure that self-evidence can still be found in the well-formed icons *regardless* of whether or not the observer in question has learned what constitutes a well-formed icon (*RSE*). The variety of iconic representation for symbolic logic I will propose is both (1) generatively assembled, object to object, and (2) once generatively assembled, self-evident in virtue of its visuospatial form alone.

Fusing both *G-RSE* and *RSE* into a unified approach, I contend, will allow us to address the *intersubjective* and *empirical* nature of crafting self-evident iconic representation for symbolic logic.

For example, say that we start out with a certain collection of objects from which we would like to generatively build our icons, say the set {Δ, ■, ⊙}. Then, suppose that we have the following icon: "■■ ⊙ ■■Δ■■■■". We then eagerly test the self-evidence of this icon with an equally eager group of participants or interlocutors. We separate our participants into two groups: Group 1 endeavors to *learn* what constitutes a well-formed icon and evaluates the self-evidence of the icon *after successfully generalizing the given patterns* while Group 2 *does not learn* what constitutes a well-formed icon and analyzes the given icon solely for representational self-evidence.

Say that we had assigned the meaning of addition to the object "⊙" and the meaning of equality to the object "Δ". And assume we had treated each "■" numerically. Thus, the icon "■■ ⊙ ■■Δ■■■■"—for Group 1—would be found to mean "$2 + 2 = 4$". Assuming clarity and precision of communication, Group 1 would arrive at a generalization of the instances of ■, ⊙ and Δ in accordance with the numerical interpretation.

Yet, consider what would happen with Group 2.

Remember that those in Group 2 have not learned anything about what makes the given icon well-formed. Thus, when they perceive the icon,

■■ ⊙ ■■Δ■■■■

they perceive an ambiguous assortment of shapes. Although it might be tempting to retort that it "should be obvious" that there is some sort of numerical connection between "■■", "■■" and "■■■■", with only one icon as example and with no other instances of the objects to generalize, there would be no way for the members of Group 2 to be sure that the icon represented "$2 + 2 = 4$" instead of "$2 \times 2 = 4$" or even some other more abstract and non-numerical statement.

Consequently, members of Group 1 could come away with the assertion that

■■ ⊙ ■■Δ■■■■

is "self-evident" (adhering to *G-RSE*). Yet members of Group 2 would not be able to assert the icon's self-evidence (as understood by *RSE*). This is an obvious dilemma. If we are seeking *maximally self-evident* iconic representation, this mismatch between *G-RSE* and *RSE* cannot do.

This is where the fusion of both *G-RSE* and *RSE* becomes crucial. Not only do we want to have our "Group 1" be able to discern *G-RSE* in the icon in question, but we also want our "Group 2" to be capable of discerning *RSE*. Moreover, if we find that our Group 2 consistently fails to discern *RSE*, we may be able to tweak our starting objects and their patterns of arrangement. We can use the dynamic nature of semiosis to our advantage.

Admitedly, however, it is not realistic to assume that *every* logical or mathematical idea can achieve *RSE*. Indeed, we can be almost certain that there are ideas whose complexity far exceeds the bounds of what is condensable into strict *RSE* (considering even a dynamically interactive and evolving computer rendering of iconic representation). As complexity increases, iconic representation will necessarily rely more on *G-RSE*. Yet, I contend that the ideas behind each of the well-known logical connectives $\{\neg, \wedge, \vee, \rightarrow, \equiv\}$ (and even *quantifiers* and *variables*) can in fact achieve *RSE*.

In addition to the "trick" of fusing *G-RSE* and *RSE*, we can also make our objects and their associations more *pictorial* in nature (and indeed this helps to fuse the two halves!). Although Lincos is a marvelously detailed and sophisticated example of self-evidence in action, Freudenthal almost exclusively relied upon letters of natural language and conventionally known symbols. This avoidance of more pictorial forms of representation no doubt inhibited Lincos from approaching *RSE*.

We have established this much: the boundaries of self-evidence can indeed be pushed. But how precisely do we do it?

11 Seeking Semiotic Self-Evidence

We can now return to C. S. Peirce's and George Spencer-Brown's projects of logical representation and analyze them in a new, Lincos-informed light. We have seen that although the thinkers' notations achieved something called *operational self-evidence* (*OSE*), they did not achieve representational self-evidence of any kind (*G-RSE* or *RSE*). To recap, we can take a look at the icons utilized by both Peirce and Spencer-Brown in addition to Freudenthal. We might as well pick an arbitrary logical connective and see what each of these thinkers has to, quite literally, *show* for himself! I pick the logical connective of *implication* ($\rightarrow$ or $\supset$):

Implication for C.S. Peirce:

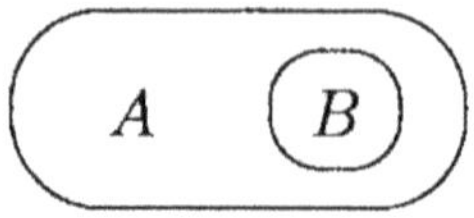

Implication for George Spencer-Brown:

$$\overline{a\,|}\;\; b$$

Implication for Hans Freudenthal:

$$\begin{aligned} \#\, a > 100\,.\!\rightarrow\!.\, a > 10\,\cdot \\ a > 1101\,.\!\rightarrow\!.\, a > 1\,\cdot \\ a > 11\,.\!\rightarrow\!.\, a > 11\; \# \end{aligned}$$

The first two examples possess *OSE*, but they do not possess *RSE*. Freudenthal's example, on the other hand, possesses solely *G-RSE*.

To see more clearly why each of the above icons fails to reach *RSE* we can take an "extraterrestrial's eye view" and consider how an alien eye would perceive the arrangement of shapes in each case.

In the case of Peirce, it might initially appear that some sort of *containment-based* notion is meant to be communicated. Our E.T. friend might get the idea that there is a set $\{A, \{B\}\}$. It is by no means clear, without having been already taught Peirce's means of representation, that the idea associated with *logical implication* is being communicated.

In the case of Spencer-Brown, it is clear that the letter "*a*" is being marked and somehow "set apart" from the letter "*b*". But it is by no means clear, without being in possession of Spencer-Brown's iconic "key", that this icon is intended to represent the idea behind "$a \supset b$."

Assuming that we have just been smacked in the face (or, more accurately, eyeballs!) with Freudenthal's arrangement of objects above, it seems we would have an even harder time deciphering what we were looking at. There seems to be no intrinsically discernible connection between the objects that imparts the idea behind logical implication.

12 A New Avenue

Now suppose that we offered the following icon to our extraterrestrial viewer:

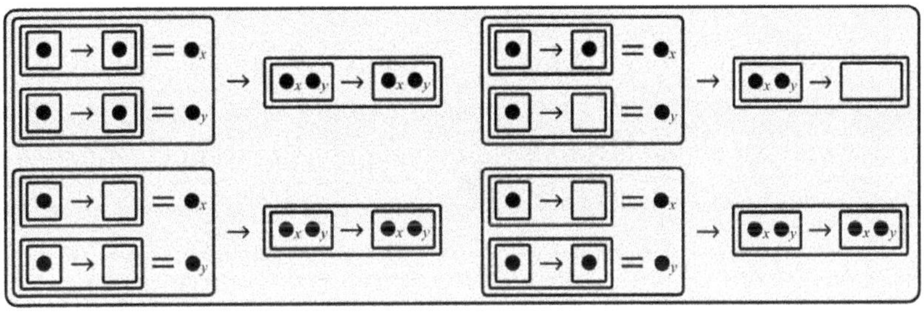

Although it may indeed be a bit of a sensory overload for our E.T. friend, it seems as though it would have a *higher probability* of recognizing that the basic idea behind logical implication were being communicated. Let us go ahead and break down this icon, object by object.

The icon above follows the intuition offered by Aristotle in his *Metaphysics*:

> To say of what is that it is not, or of what is not that it is, is false, while to say of what is that it is, and of what is not that it is not, is true. (Metaphysics 1011b25)

That is, logic is fundamentally based on the *presence* and *absence* of generic things, their relative *existence* and *non-existence.*

In the above icon there are only four breeds of object: *boxes*, *circles*, *arrows* and *equals* objects.

In some instances we have a box with a circle inside of it:

In other cases, we have a box without anything inside of it:

Sometimes we have our circles stand alone, without any box around them, like:

and like:

There are also cases of two boxed circles being related by arrows, like:

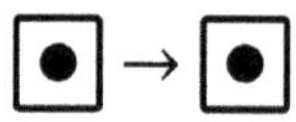

and like:

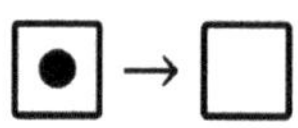

Both of the above patterns are icons representing Aristotle's criteria for truth and falsity. The first pattern "[•] → [•]" represents one half of Aristotle's criterion for truth: *to say of what is that it is.* And the second pattern "[•] → [ ]" represents one half of his criterion for falsity: *to say of what is that it is not.*

As we are teaching the well-formedness of such icons to our viewers (extraterrestrials or simply eccentric mathematicians!) we might stress *both halves* of Aristotle's criteria for truth. For instance, we might offer our viewer the following arrangement of objects (we can borrow Freudenthal's "#" marks to illustrate the beginning and end of our iconic display):

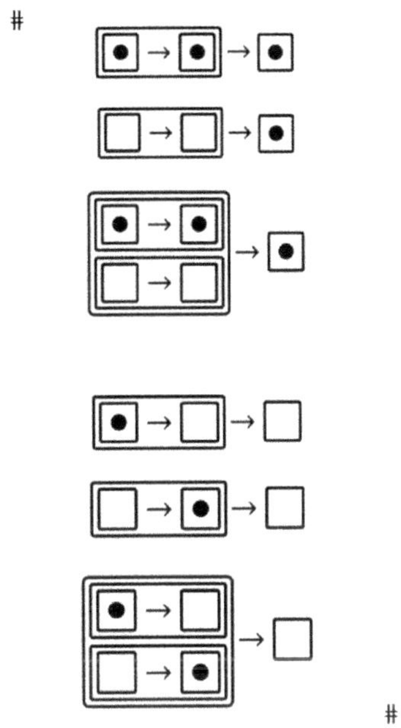

These icons seem to impart an intuitive idea of logical truth and falsity.

If we were to continue following Freudenthal's Lincos cue, we might offer the following arrangement of objects in the beginning of the formation of our icons:

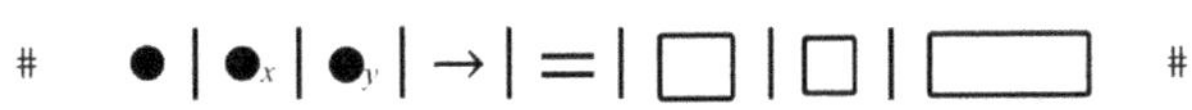

Having already offered icons imparting the intuition of Aristotle's defintion of logic, we can now address some other parts of the icon representing logical implication:

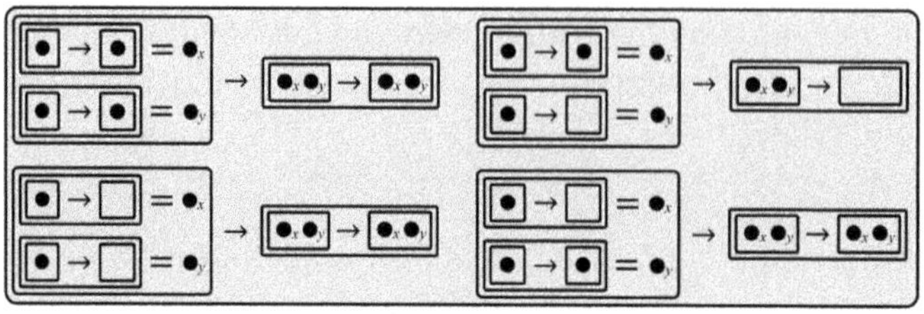

We must admit that most of us humans have an unfair advantage when it comes to discerning the meaning of the "=" object. How can we be so sure that an extraterrestrial eye could discern the intended meaning of "="? Thankfully, I have a straightforward approach in mind to ensure that the observer has a reasonable opportunity to learn the meaning of "=".

Conveniently enough, the arrangement of objects which presents the meaning of "=" also presents the meanings of the two dots subscripted with "x" and "y".

For example, we might offer a generic observer the following arrangement of objects implicating the dot of subscript "x":

#

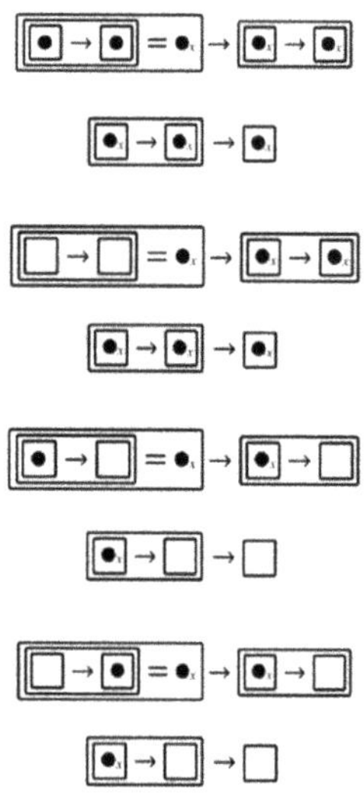

#

And following this, we might show the very same arrangement, except now with the dot of subscript "y":

#

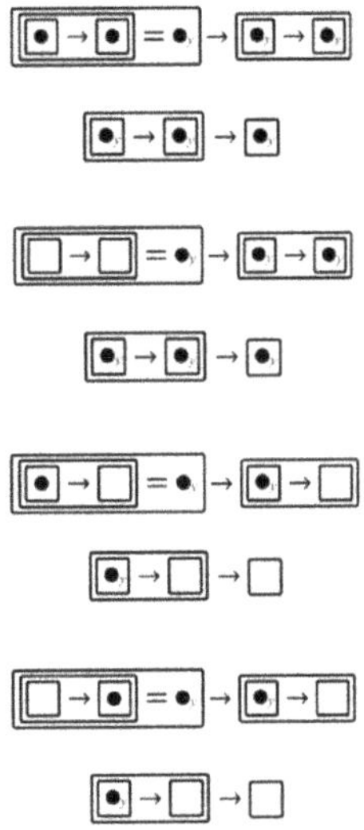

#

A certain pattern that we draw on is the following:

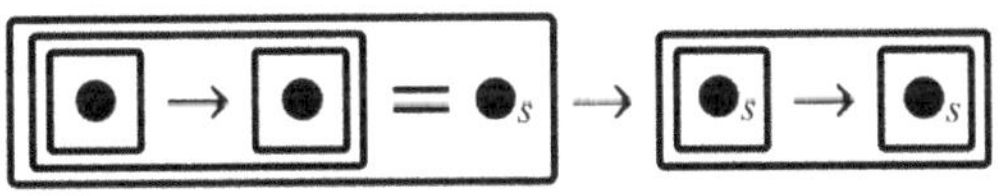

This icon has three key components. First, there is the portion that represents Aristotle's intuition of logic (in this case we are showing that what is, *is*):

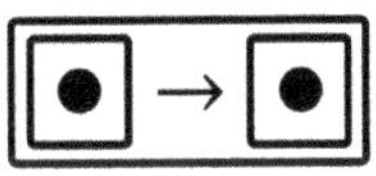

Second, there is a portion illustrating that we are "setting" or "initializing" some new object to the value true:

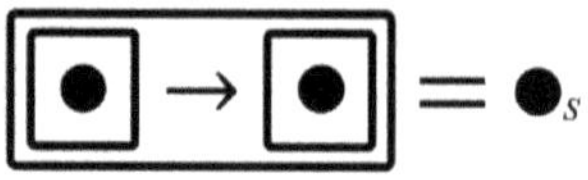

Finally, we have the representation that everything to the left of the arrow results in this new object itself having the value true (or that it adheres to Aristotle's "what is, *is*"):

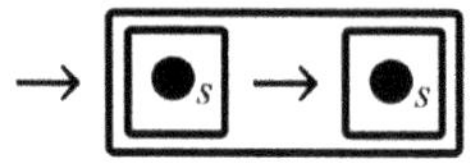

In the "program text" or arrangement of objects shown above, we show all four possible arrangements of truth value: (1) to say of what is that it is, (2) to say of what is that it is not, (3) to say of what is not that it is and (4) to say of what is not that it is not. The icons adhering to (1) and (4) represent *truth*, while those evincing (2) and (3) illustrate *falsity*.

This basic pattern is shown in conjunction with the following different pattern:

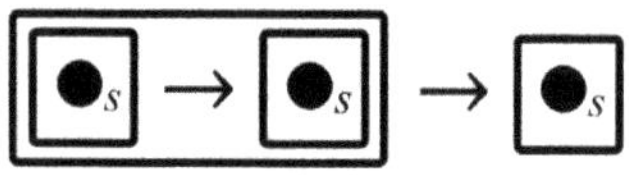

All this pattern shows is that the truth value of the subscripted circle object *also* conforms to the Aristotelian maxims. The pattern above illustrates *inherited truth*, while the following pattern illustrates *inherited falsity*:

Having explained the notions of truth and falsity, in conjunction with the "=" object, the arrow object, and the subscripted circles, we have only one last component of the icon for logical implication to explain. Let us take one more glance at the icon:

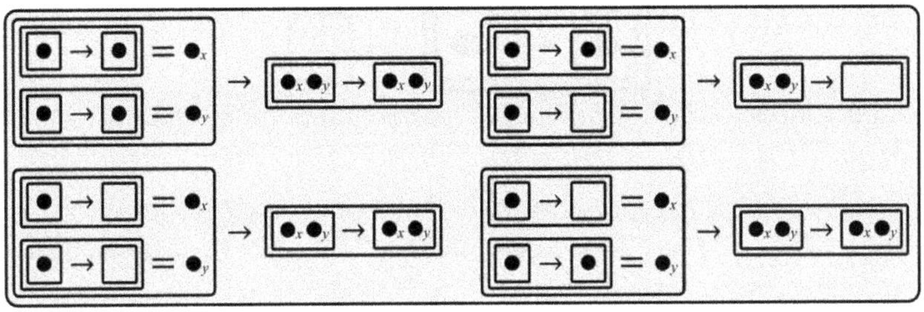

An interesting part of this icon is the grouping of the two subscripted circles in one box like so:

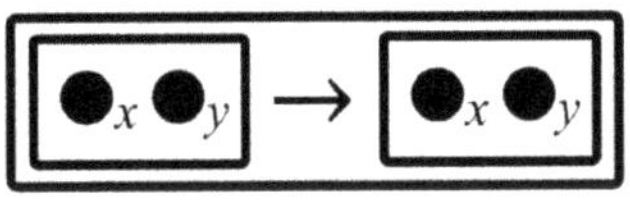

One might be tempted to ask why there is a need for such composite objects at all. Why can we not have one circle in each box? The simplest answer is that these composite circles are necessary because the implication connective is 2-ary and not 1-ary. It might be helpful to consider what an icon for the 1-ary logical connective "¬" might look like:

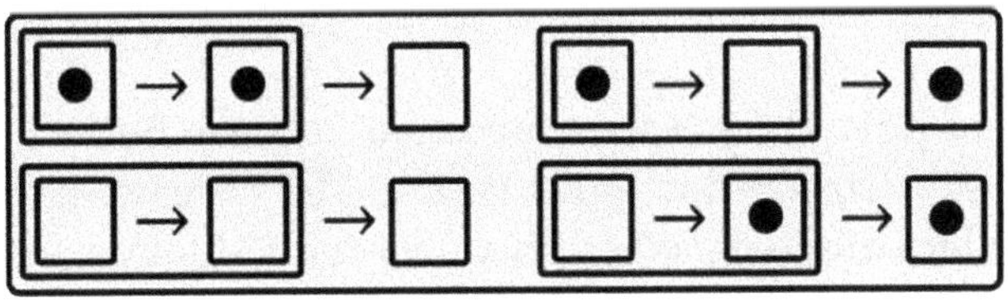

In the case of a 1-ary connective like negation, there are no composite objects (like the two subscripted circles in the 2-ary implication icon) necessary. Because of the 1-ary nature of negation, simple associations like the following do the trick:

Yet, in the case of every 2-ary logical connective, we must have associations like the following:

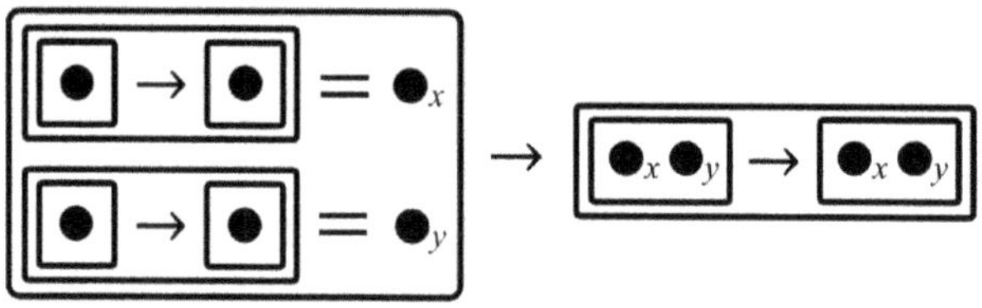

We could, in theory, also have an arrangement like so:

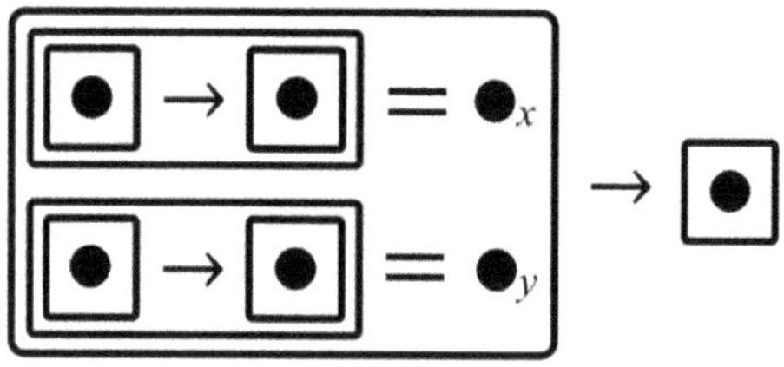

However, this would leave the precise connection between both the "x" subscripted dot and the "y" subscripted dot quite vague. To keep with the intuition that propositional variables are assigned truth values individually and that truth values of sentences are determined by composition of these valued variables, it seems that the composite circles in the icon are helpful visual aids.

With the implication connective "⊃" in the bag, we can dive into the other four main connectives and display each of our icons in tandem with one another. In addition I will briefly show how this approach allows one to self-evidently verify the propositional tautologies.

13 Self-Evident Connectives

¬

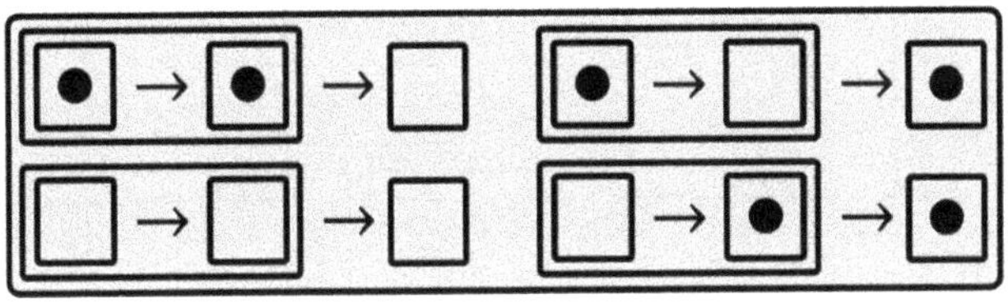

∧

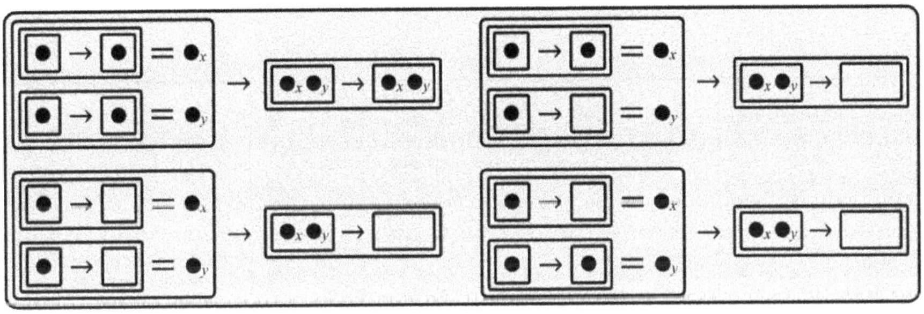

∨

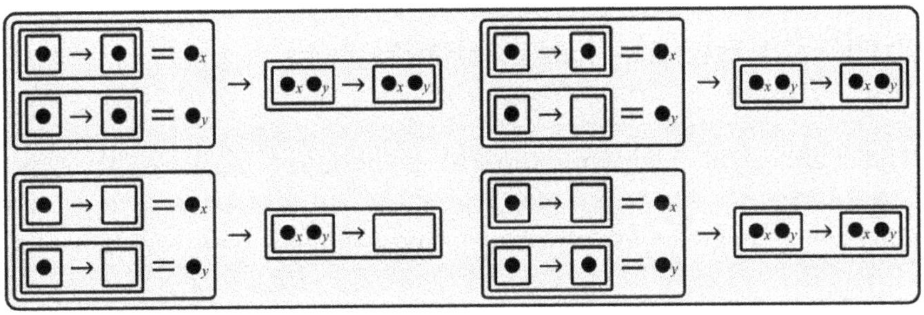

$\supset$

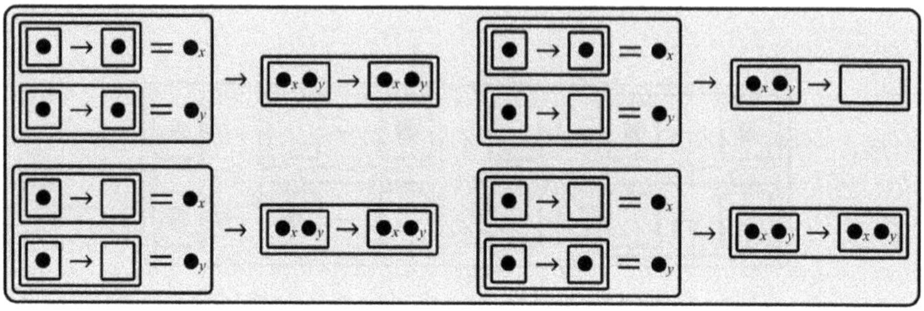

$\equiv$

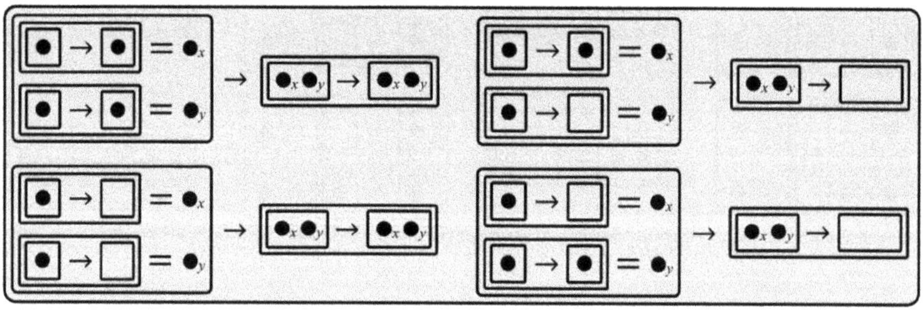

14 Self-Evident Tautologies

I will now offer a method of verifying the status of modus ponens (MP) as a tautology. Firstly, I will introduce some ideas for representation and explain them. Following this, I will offer a Lincos-style "program" that amounts to a self-evident proof of MP.

How shall we iconically prove/represent the tautology "$(p \wedge (p \supset q)) \supset q$"? We can start with some icons invoking Aristotle's notion of logic, as before:

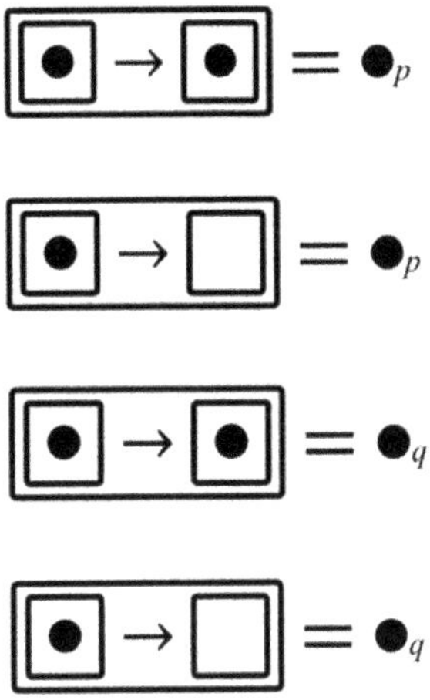

What we can do is illustrate a "case-by-case" evaluation of "$(p \wedge (p \supset q)) \supset q$", starting with each of the truth-valued combinations of p and q. For example, we can start with the case of p as T and q as T:

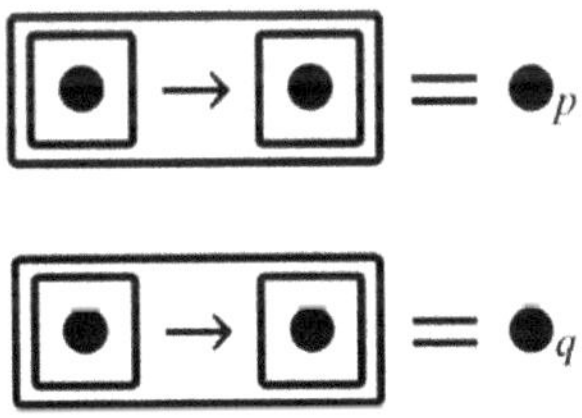

In the context of these two icons, we can then examine the respective structures of the icon for "$\wedge$" and that for "$\supset$":

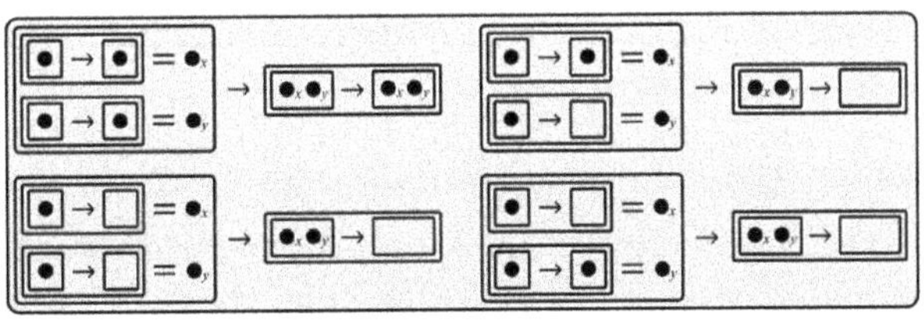

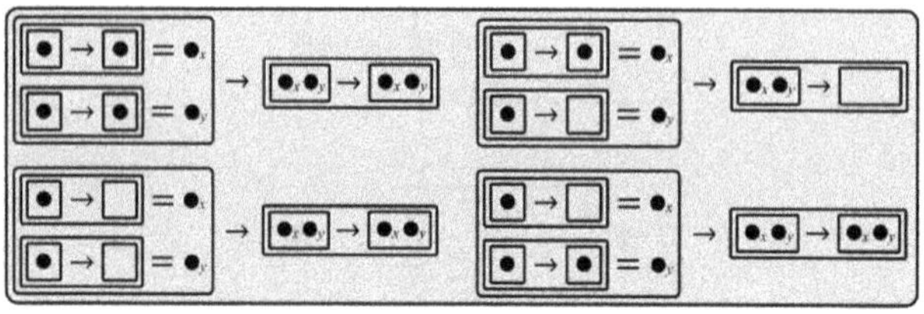

Let us assume that we have assumed the truth of p and the truth of q as in the above example:

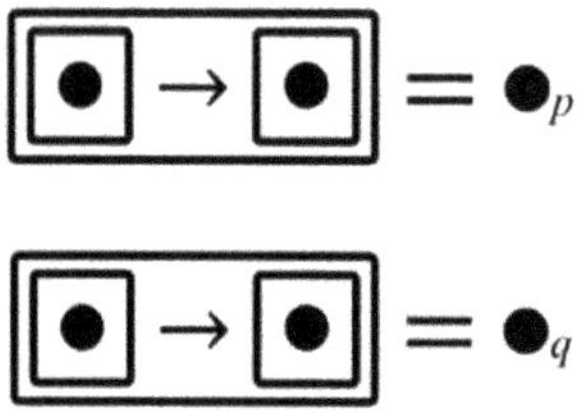

To self-evidently illustrate "$(p \wedge (p \supset q)) \supset q$", much like we do symbolically, we start with the nested expression "$(p \wedge (p \supset q))$" and move from left to right. An easy way for us to illustrate the "searching process" for corresponding truth values is through shading.

For example:

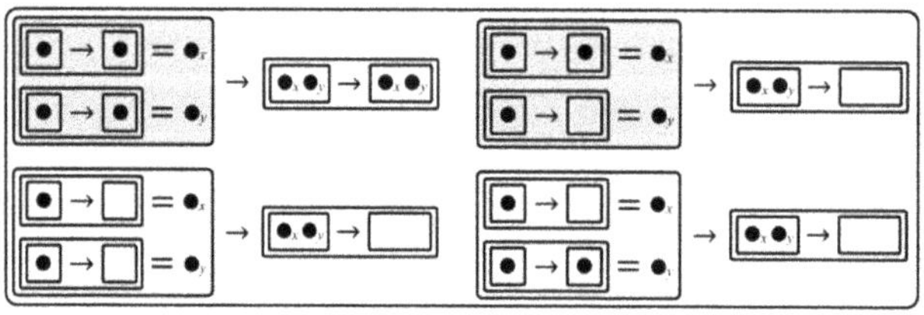

This icon offers to intuition that certain possibilities corresponding to a T-valued p have been provisionally singled out. Notice that there is no shading yet in the composite boxes (those containing both subscripted circles). This shading of the icon corresponding to "∧" represents the left-most portion of the expression "$(p \wedge (p \supset q)) \supset q$". In this sense, we might rather sloppily say that the above icon represents "$(p \wedge (p \supset q))$" or that the icon represents the truth assertion of p (and only p) *in the context of* a generic statement "$(p \wedge q)$".

After this icon, we can move on to the expression "$(p \supset q)$". We switch to the iconic template corresponding to the "⊃" connective, hold the truth values (T and T) that we assumed for both p and q steady and shade those boxes which correspond to these truth values:

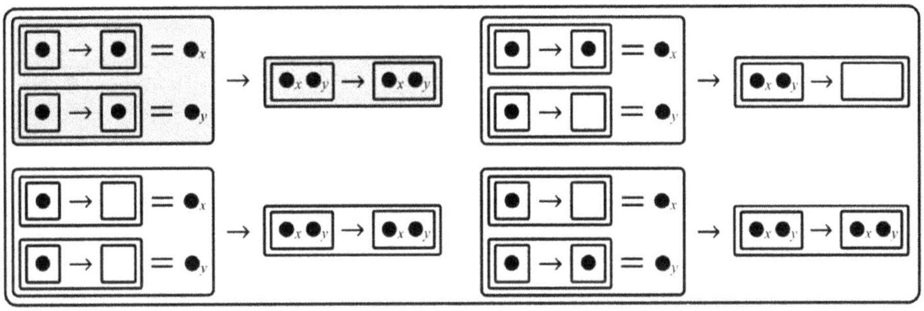

We then notice that assuming both p and q true leads to a true result in the context of the "⊃" icon. Having evaluated the "$(p \supset q)$" component of the expression "$(p \wedge (p \supset q)) \supset q$" to T—and recalling that we have assumed p as true—we can then move back to the "∧" icon and shade the boxes corresponding to their truth values:

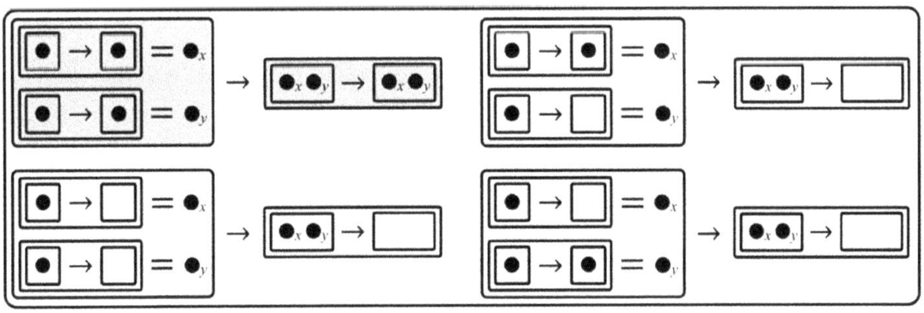

We then see that the "$(p \wedge (p \supset q))$" part of the tautology "$(p \wedge (p \supset q)) \supset q$" evaluates to true. The last step is to switch to the iconic template corresponding to the "⊃"

connective once more, hold the truth values (T and T) that we assumed for both p and q steady, remember the composite truth values we derived and shade those boxes which correspond to the truth values we have assumed:

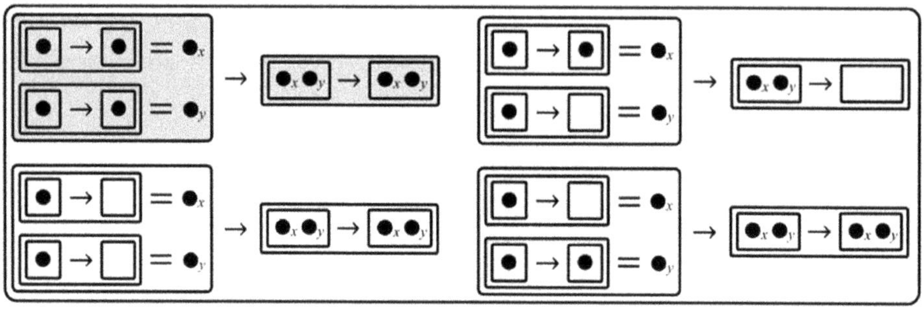

This progression from icon to icon represents a completion of the evaluation of "$(p \wedge (p \supset q)) \supset q$" when assuming the truth of both p and q. To verify the tautology itself, all that is needed is to continue the same method for all possible truth values.

I will now recap each of the iconic methods I have unveiled. I will offer a series of "program texts" in the style of Hans Freudenthal's Lincos, starting from mere objects and generatively assembling into iconic representation of propositional logic and modus ponens. Note that while above I have used the traditional "p" and "q" for sentential variables, in the program texts I will use only x and y for continuity.

15 Hey, E.T., This is Logic

● | $●_x$ | $●_y$ | → | = | □ | □ | ▭ | ¬ | ∧ | ∨ | ⊃ | ≡ | (|)

● | $●_x$ | $●_y$ | → | = | □ | □ | ▭ | ¬ | ∧ | ∨ | ⊃ | ≡ | (|)

#

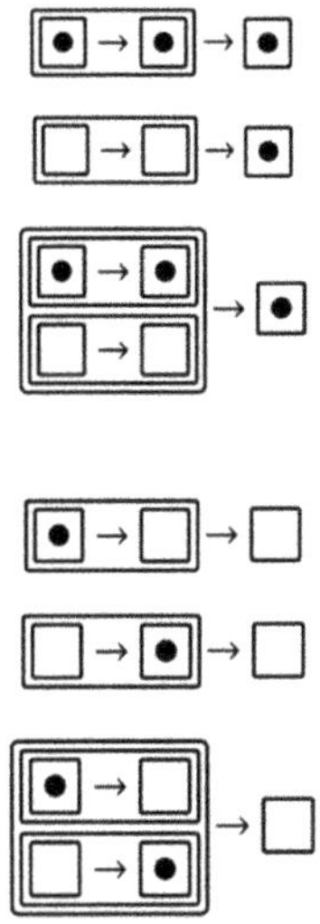

#

#

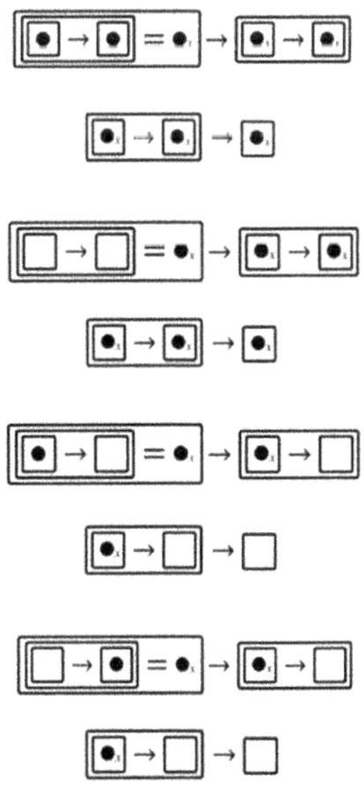

#

#

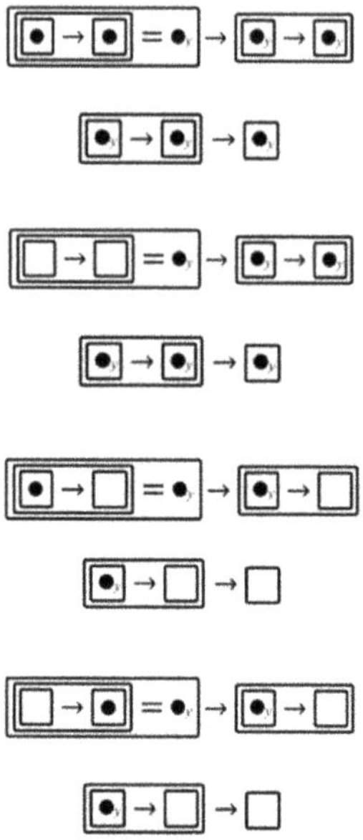

#

#

¬

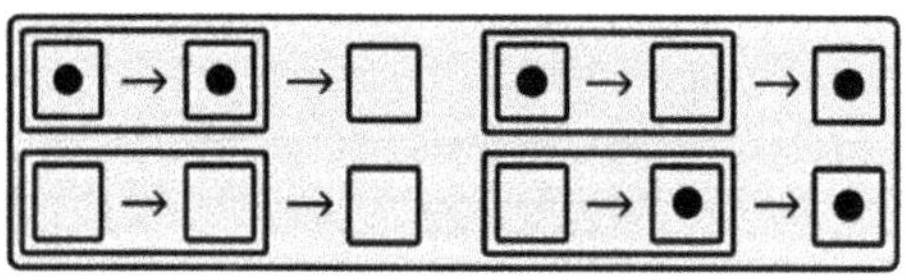

∧

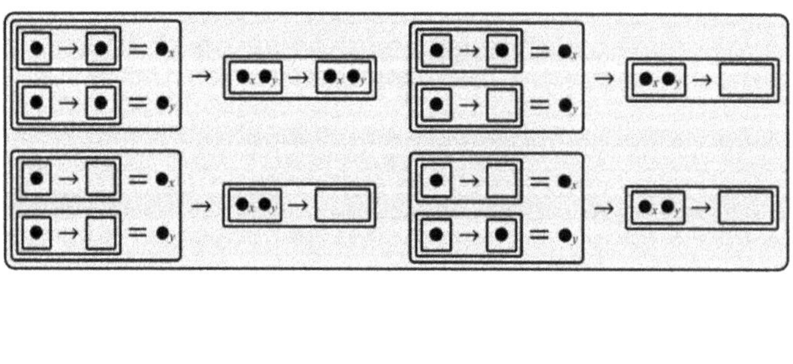

$\vee$

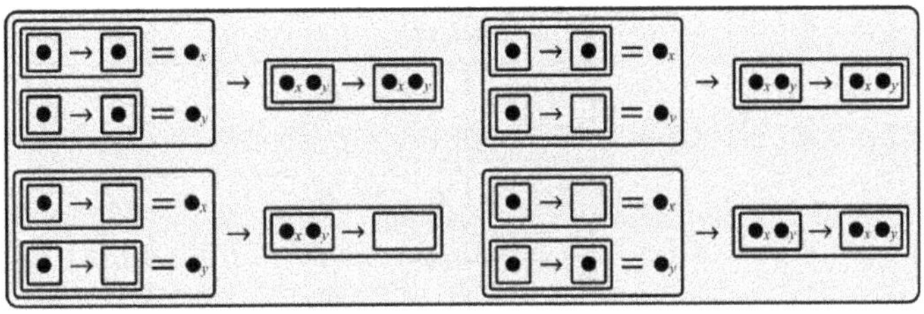

$\supset$

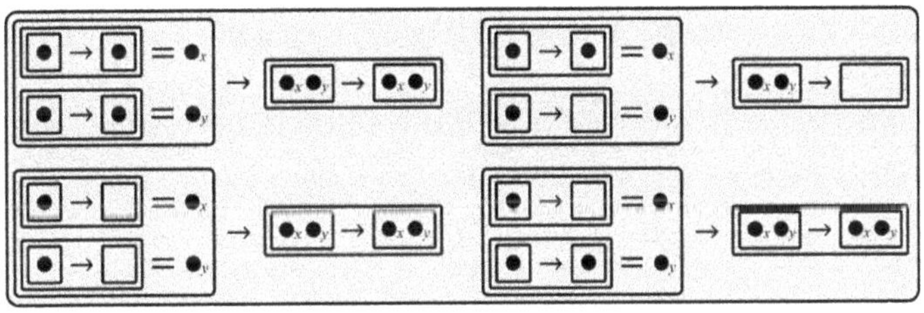

$\equiv$

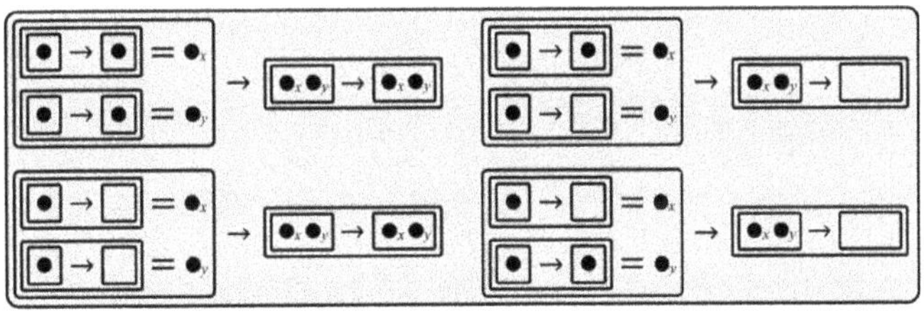

#

#

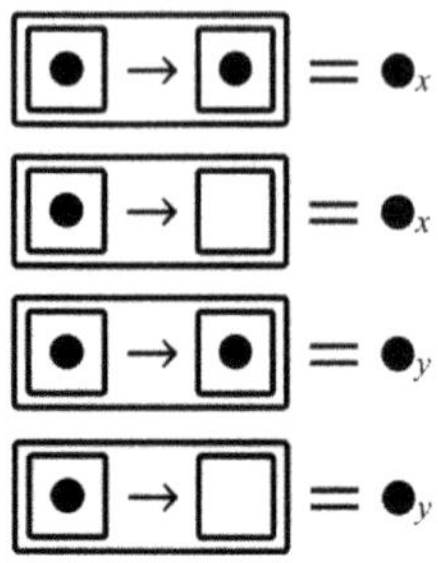

#

#

$$(\bullet_x \wedge (\bullet_x \supset \bullet_y)) \supset \bullet_y$$

#

#

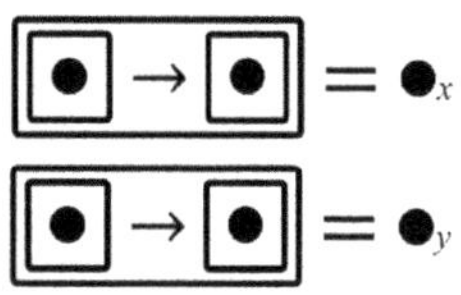

$\wedge$

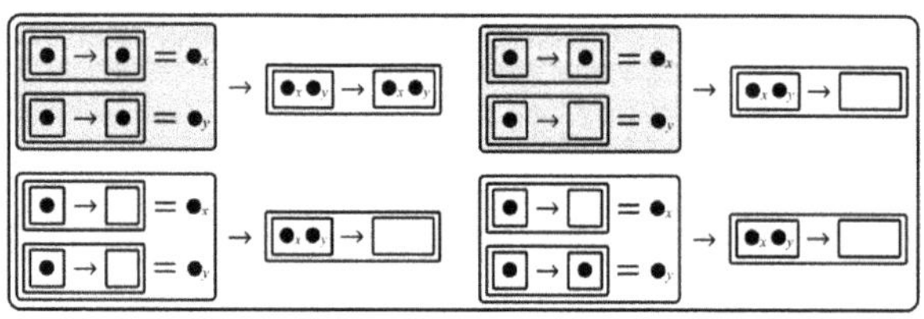

$\supset$

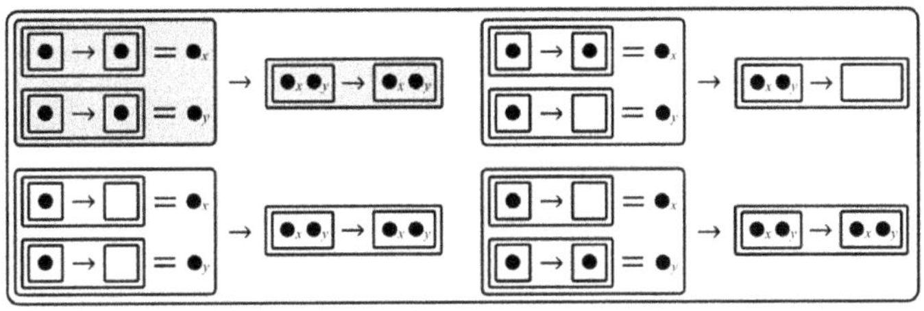

$\wedge$

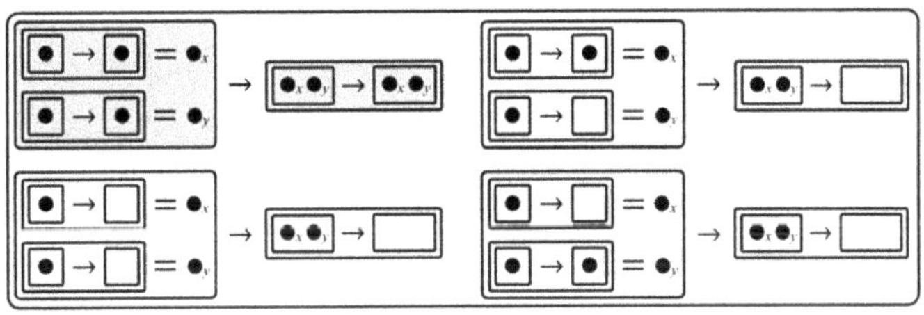

$\supset$

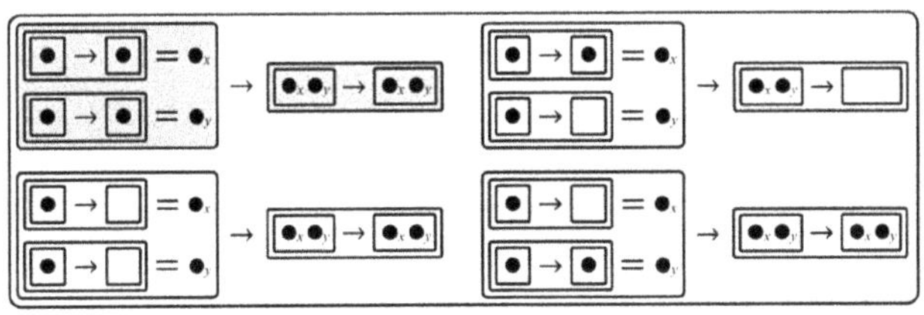

#

#

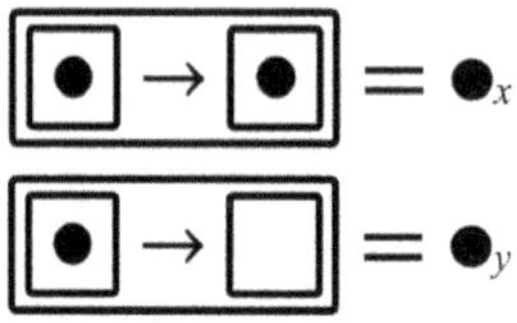

∧

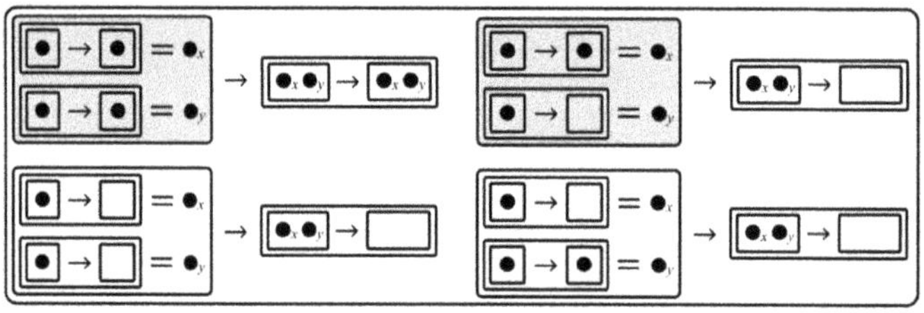

⊃

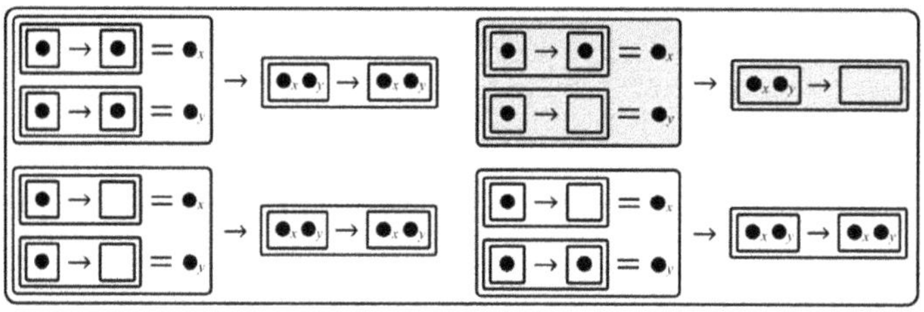

∧

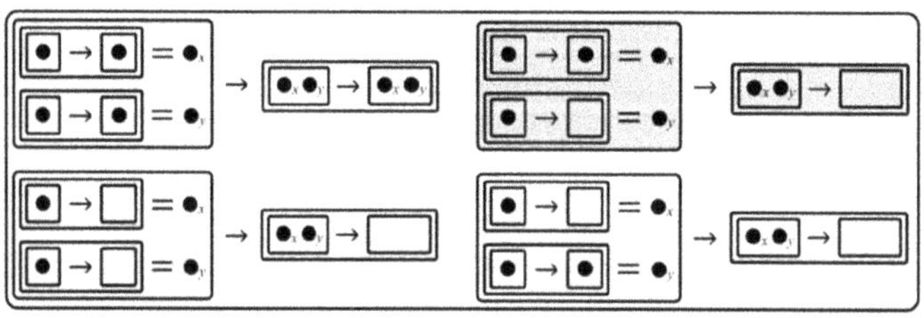

⊃

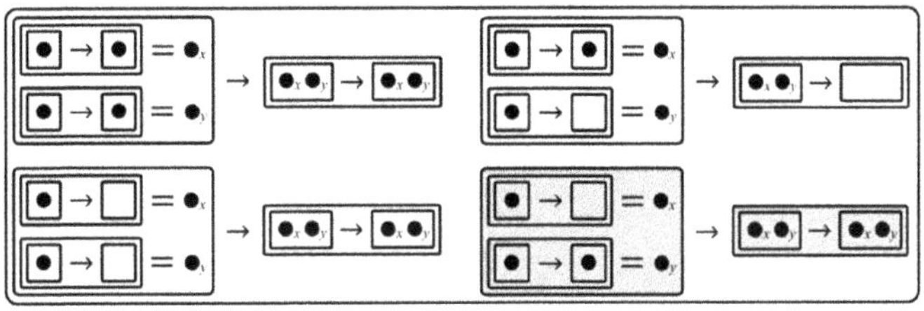

#

#

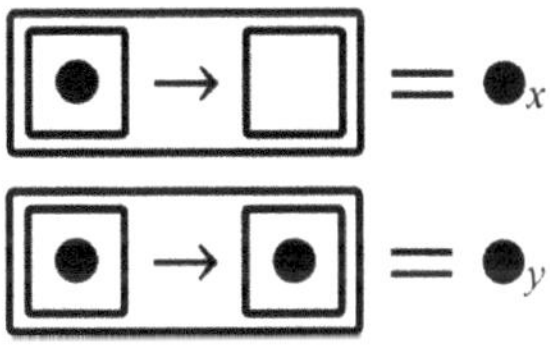

∧

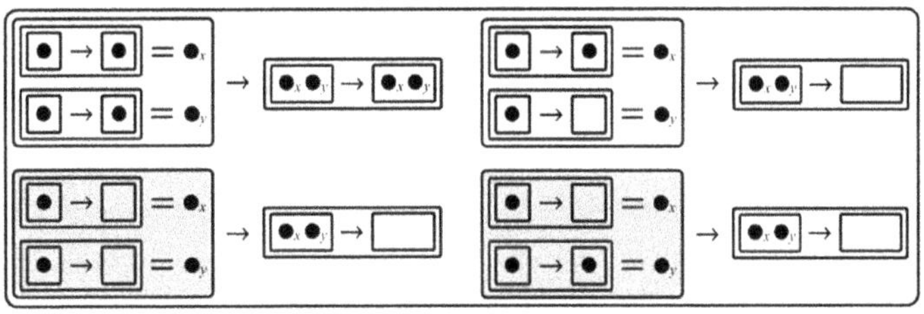

$\supset$

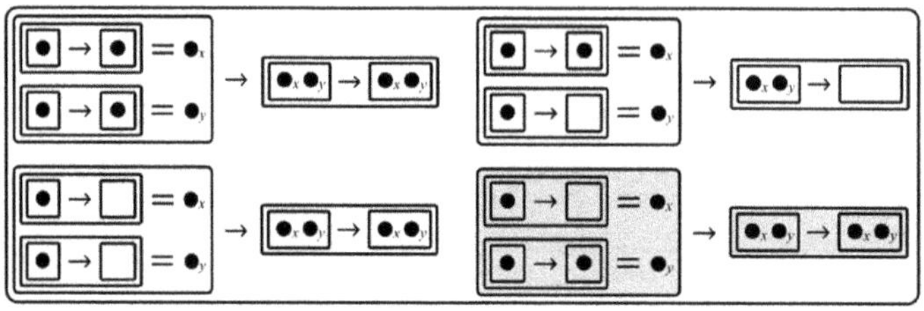

$\wedge$

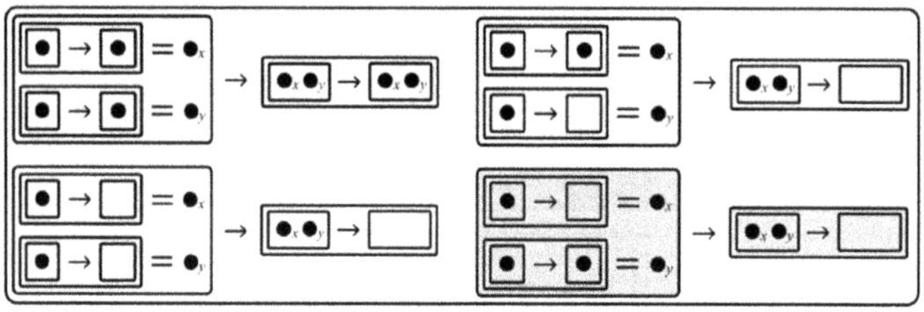

$\supset$

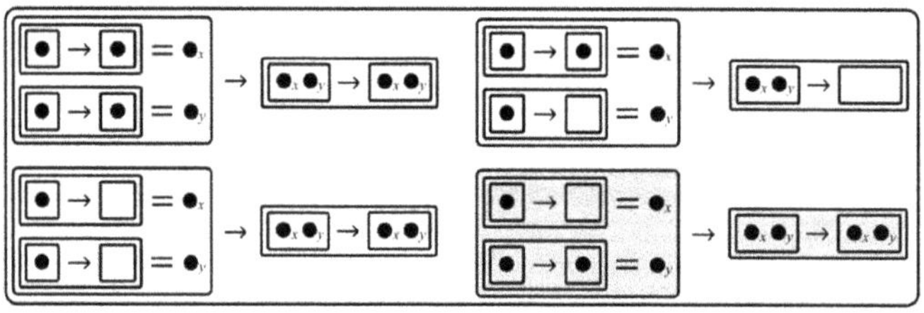

#

#

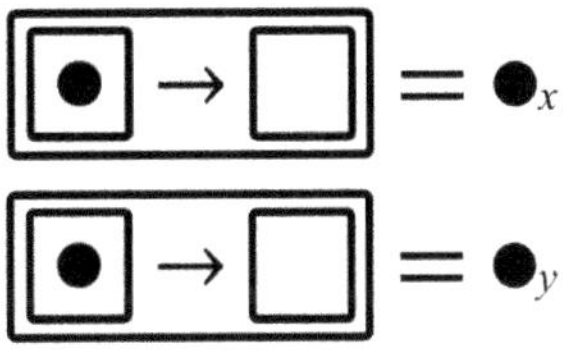

∧

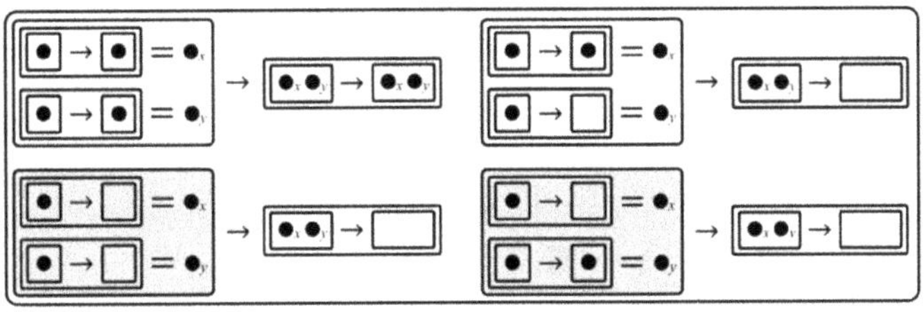

⊃

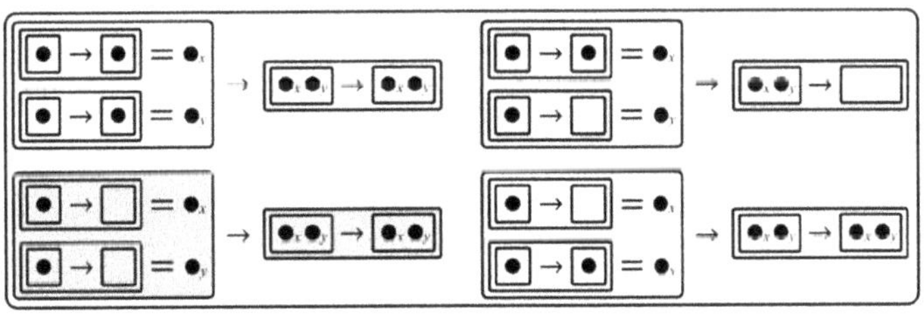

∧

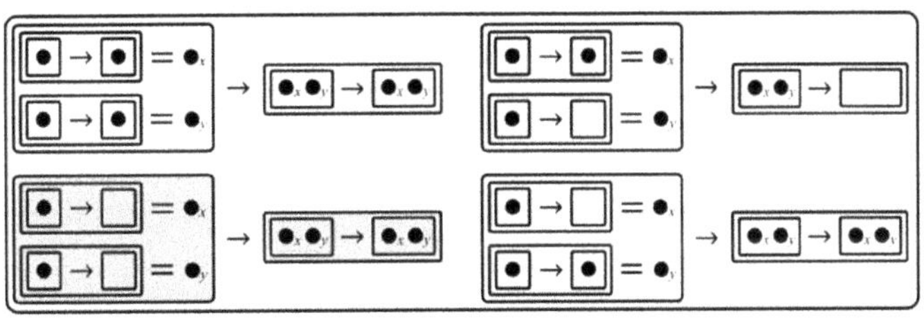

$\supset$

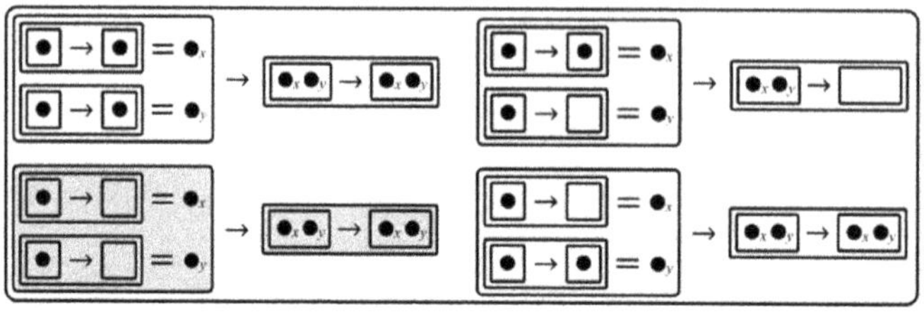

♯

♯

$$(\bullet_x \wedge (\bullet_x \supset \bullet_y)) \supset \bullet_y$$

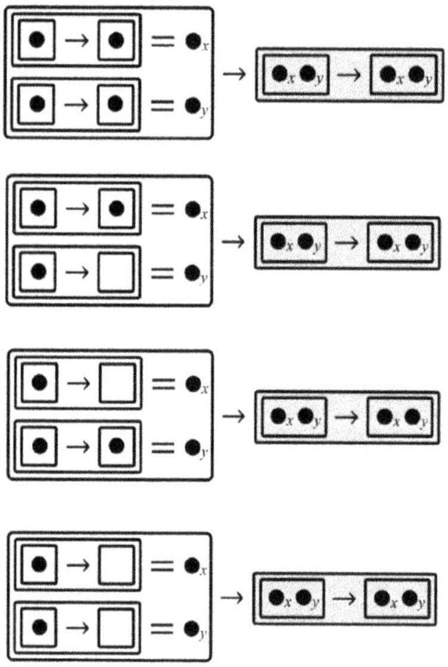

♯

16 Visions of the Future

Perhaps this form of iconic representation is simply a cute, slightly clever rendering and nothing incredibly special. But I would like to maintain at least some conviction in its capacity to represent and shed light on logical and mathematical notions *much more complex* than propositional logic. Specifically, I would like to briefly discuss my approach as it relates to the *computer visualization of mathematics.*

Starting in 1988 with Stephen Wolfram's *Mathematica*, mathematicians' toolkits expanded beyond their pens and paper, their minds and their chalkboards. Computer software such as *Mathematica, Maple, MATLAB* and *Penrose* allow mathematicians to dynamically plot, render and model mathematical structures in straightforward, graphical ways. Those mathematicians not endowed with spatial acumen need not fret any longer; they can simply plug a specification into such software and see a high-resolution, rotatable geometric structure right before their eyes. And by no means are such graphical visualization engines confined to mathematical objects of the geometric kind. Many such engines allow the user to render objects like *sets* or *groups.* Below is an example of a rendering of set-based containment in Penrose (Figure from Ye et al., 2020):

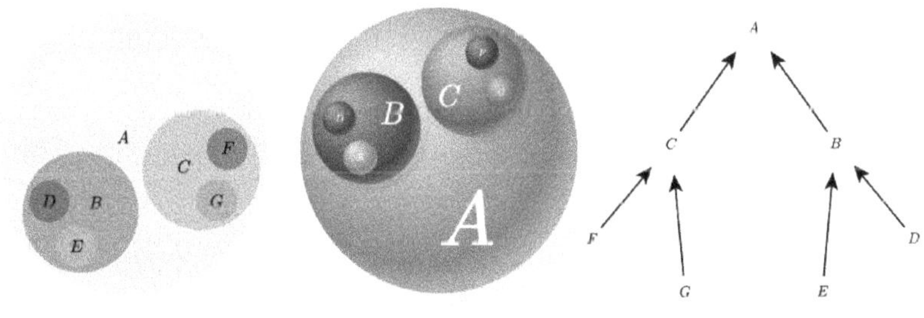

Fig. 18. Here, some SUBSTANCE code is used to specify set relationships. Different STYLE programs not only tweak the visual style (*e.g.*, flat vs. shaded disks), but allow one to use a completely different visual representation (*e.g.*, a tree showing set inclusions). `Sets.sty` above describes the flat disk style.

Such software is profoundly useful and continues to open up new horizons for mathematical interaction and progression. Yet, we cannot help but be reminded of the criterion of *representational self-evidence* that we have encountered above:

RSE *Call a given group of* icons *representationally self-evident if, when one is presented with a well-formed arrangement of these icons, it is immediate from the visuospatial form alone—without any symbolic context necessary—which idea(s) are intended to be represented.*

Although the graphical renderings of such engines as *Penrose* and *Maple* serve as tools to aid mathematicians, the graphics themselves—excluding perhaps set-theoretic graphics, as the notion of containment is a straightforward one—rely upon their *surrounding context* for interpretation.

Not only do these graphics fail to achieve *RSE*, but they also manage to say little about *the semiotic process of the theorist himself/herself* and how this relates to the mathematical entities being modeled.

Imagine an entirely new type of graphics engine which offers the following:

- The graphics of the engine strive to achieve *RSE* (taking into consideration that *RSE* is often *relative to accumulated knowledge*),
- The graphics of the engine are *generated* (via code) *by the user from scratch, via an arrangement of simple objects* if desired, in the vein of *Lincos* and the breed of iconic representation I have offered,
- The ability to *visualize* not only geometric and algebraic mathematics but such fields as *proof theory*, *model theory* and *number theory* as well,
- The ability to *dynamically track the path of mathematical derivation* (considering that both pre-symbolic and symbolic signs can be programmatically generated by the user and this process of generation itself can be visualized as an interactive diagram),
- The ability to take both an "ant's eye view" and a "bird's eye view" of mathematics, starting from the pre-symbolic level and generating relevant indices, icons and symbols, while also being able to bring such semiotic process to bear on the process of mathematical reasoning as a whole,
- The ability to distill disparate proofs of differing branches of mathematics into similar "threads" of semiotic process involving the generation of signs as described above.

Certainly, such a graphics engine is far off. It first must be demonstrated that not only propositional logic but *predicate logic* and *set theory* as well can be self-evidently generated without any symbolic context. I contend that this can indeed be done and plan to write a subsequent paper detailing the approach (in addition to discussing the process of generating not only icons but *symbols* themselves). But such prospects of an entirely new mathematical graphics engine are exciting nonetheless.

In the past century, both mathematics and academia at large have grown increasingly sectarian. Bright minds often manage to speak past each other entirely. An algebraic geometer and a model theorist likely have little to converge upon. A microbiologist and a sociologist may just manage to chat about politics and the weather. But, at rock bottom, *each of us swims through the same semiotic stream.* If we can manage to wake up, look around us and realize our similarities, some quite marvelous things could be around the corner.

On this note, we might as well close with a slightly nuanced excerpt of David Foster Wallace's wildly famous graduation speech:

> There are these two [grad students walking] along, and they happen to meet an [older academic walking] the other way, who nods at them and says, "Morning, boys. How's the [semiotic process]?" And the two young [grad students walk] on for a bit, and then eventually one of them looks over at the other and goes, "What the hell is [a semiotic process]?"

References

[1] Allwein, Gerard, and Jon Barwise, eds. 1996. *Logical Reasoning with Diagrams.* Studies in Logic and Computation. New York: Oxford University Press.

[2] Angelo, Joseph A. 2007. *Life in the Universe.* New York: Facts on File.

[3] Aristotle. 1989. *Metaphysics.* Translated by Hugh Tredennick. Loeb Classical Library 271–287. Cambridge, MA: Harvard University Press. First printed 1933–1935

[4] Boole, George. 2009. *An Investigation of the Laws of Thought.* Cambridge: Cambridge University Press. First printed 1854.

[5] Bricken, William. 2018. *Iconic Arithmetic: The Design of Mathematics for Human Understanding.* Snohomish, WA: Unary Press.

[6] Freudenthal, Hans. 1960. *Lincos: Design of a Language for Cosmic Intercourse.* Amsterdam: North-Holland Publishing Company.

[7] Langan, Christopher M. 2017. "An Introduction to Mathematical Metaphysics." *Cosmos and History: The Journal of Natural and Social Philosophy* 13 (2): 312–330.

[8] Langan, Christopher M. 2018. "The Metaformal System: Completing the Theory of Language." *Cosmos and History: The Journal of Natural and Social Philosophy* 14 (2): 207–227.

[9] Peirce, Charles Sanders, and Matthew E. Moore, eds. 2010. *Philosophy of Mathematics: Selected Writings.* Bloomington: Indiana University Press.

[10] Peirce, Charles Sanders, ca. 1895. *On the logic of quantity.* Autograph manuscript.

[11] Peirce, Charles Sanders, and Peirce Edition Project, eds. 1998. *The Essential Peirce.* Vol. 2. Bloomington: Indiana University Press.

[12] Roberts, Don D. 1973. *The Existential Graphs of Charles S. Peirce.* The Hague: Mouton.

[13] Spencer-Brown, George. 2011. *Laws of Form.* Leipzig: Bohmeier Verlag. First published 1969 by George Allen & Unwin.

[14] Steele, Bill. 1999. "It's the 25th Anniversary of Earth's First Attempt to Phone E.T." Cornell Chronicle, November 12, 1999.

[15] Wallace, David Foster. 2009. *This Is Water: Some Thoughts, Delivered on a Significant Occasion about Living a Compassionate Life.* New York: Little, Brown.

[16] Ye, Katherine, Wode Ni, Max Krieger, Dor Ma'ayan, Jenna Wise, Jonathan Aldrich, Joshua Sunshine, Keenan Crane. 2020. "Penrose: From Mathematical Notation to Beautiful Diagrams." *ACM Transactions on Graphics* 39. https://doi.org/10.1145/3386569.3392375.

Towards A New Monadology

Dirk Baecker
Zeppelin University, Friedrichshafen, Germany
baecker@mac.com

Abstract

Monads are the elementary operations of the universe. They not only communicate; they are communication. In Leibniz's monadology, it is God who pre-establishes their communication. A new monadology views monads in terms of Spencer-Brown's forms. Forms communicate by re-entering meta communication into communication. A calculus of communication proceeds by calculating forms into and out of their own medium. The paper thus refers to ancient thinking not of unity (Greek *hen*) but of units or single entities (Greek *monades*) from Pythagoras to Leibniz and links them to questions of language and communication (or *lógos*). Leibniz's reference to the communication with God as an explanation of the condition of the possibility of "windowless" monads' communication with each other is replaced by the idea that monads are at once their operation and the medium of their operation, so that they double as units and context, environment and condition of those units. In a sociological perspective, monads communicate by being communication. The paper is written in the style not only of Leibniz's monadology but also Ludwig Wittgenstein's *Tractatus Logico-Philosophicus* which has rightly been called "the world's longest koan" (Pearce and Cronen 1980, 67).

An earlier version of the paper has been given as a talk at the *Laws of Form* Conference 2024, Liverpool, UK, August 7—10, 2024. English language editing by DeepL Writing. Helpful comments by J. M. Flagg and two anonymous referees.

1 Monads

1. Monads are units of self-observation. Everything they perceive and know they perceive by observing their own states.

2. Classic Greek philosophy regarded monads (from Greek *monás*, unity as a single entity, in contrast to Greek *hen*, unity of the all-encompassing One) as entities maintaining themselves in relation to the void of the universe. They are determined from the relationship between the unbounded, the boundary, and their connection (*apeiron*, *peras*, *harmonia*). Pythagoras, Fr. 26f

3. Plato speaks of the bounded in relation to the unbounded. One may think of an indeterminate duality (*ahóristos dyás*) that describes this relationship, or rather to consider the mixture (*meixis*) and its cause (*aitía*), which combines the unlimited (*apeiron*) and the limited (*peras*) into the community of the many with the one. Plato, Sophist, 253d–3; Philebus, 17a, 23c–d and 26d

4. It is difficult, but promising, to take a path to the determination of the existent that does not allow itself to be discouraged by the One that is also the Many and the Many that is also One, even if one is repeatedly tempted to stop at the One or the Many. Plato, Philebos, 16c–e

5. Every beginning combines excess (*hyperoché*) with defect (*elleipsis*). The four causes of *causa formalis, causa materialis, causa efficiens* and *causa finalis* describe how excess overcomes defect and how defect informs excess. "And when we speak like this, we think we have given the cause." Aristotle, Physics, 189b 12–13; Aristotle, Metaphysics, 983a 25–33 and 1013a 23–35, quote 35–36; Wieland 1962; Weiner 2016; Luhmann 1990, 81

6. Monads are always in motion. They fall through space, but due to their unpredictable drift (Greek *parénklisis*, Lat. *clinamen*) they collide and form the abundance of phenomena. Collision means that new monads form for which all present statements hold. Leucippus, Fr. 3f.; Democritus, Fr. 49–55; Epicurus, Her. 43; Lucretius, De Rerum Natura, Book 2, 217–293

2 Operations

7. Leibniz, in his *Monadology* (1714), considered monads to be single beings, closed in on themselves, having "no windows", each reflecting the universe

from its peculiar and possibly distorted (but who is there to tell) perspective. Leibniz 1714, §§ 7, 18, 56—50

8. For Leibniz, there is no whole of which monads would be parts, even if there is God, the super-monad, who is their "sufficient cause". Leibniz 1714, §§ 37–45

9. Monads are not parts of a whole but what makes divisibility possible. They describe the possibility of operations, a continuum of discontinuities. Hecht 1998, 70; Cheung 2009

10. A monad is like a fold. It closes in on itself and remains open at the same time. It is an opening up to the inside as Leibniz discovered when he studied microbes and refractions with the new instruments and methods of the microscope and optics. Leibniz 1714, §§ 61–71; Deleuze 1993; Schuller and Schmidt 2003, 50ff.

3 Closure

11. Closure means "freedom from casual dependence", "isolation within the universe". Monads develop perceptions which start confused but get clarity when developing memory and consciousness. Eventually monads reach insights into architectural samples of the universe. They begin to understand themselves, differentiate among themselves, and consider the universe as just another monad. Leibniz 1714, §§ 21–33, 60, 82–84; quotes: Whitehead 1967, 46

12. Closure requires calculus, which is the discrete (or 'digital') calculation of an analogue reality that is considered to run continuously alongside it. Cuntz 2022

13. Closure is an operation which at the same time is an opening to counter the entity (*ordo inversus*, counter-joint and counter-direction). Novalis 1795/96, §§ 278–292; Mahnke 1937

4 Crosses

14. Monads are crosses that indicate the two sides of a distinction. These indications are at best hints and clues, or constructions that tell more about them than about the states they indicate. They indicate implicit contexts, yet

unwritten crosses, which call forth further hints and clues. Spencer-Brown 1969; Varga von Kibéd and Matzka 1993

15. Since monads are crosses, they are also observers. They produce themselves by drawing a distinction. Since they are distinctions indicated by another observer (viz., us) we are dealing with second-order observation from the very beginning and are already talking about ourselves as well. Luhmann 1990

16. There is a void indicated as the outside of the distinction. While reflecting the universe, the monad is empty. It is a function of itself, searching from the very beginning for its saturation or supplement. Frege 1891; Scheier 2016

17. Monads are not products of *creatio ex nihilo*. They appear in the middle or at the periphery of everything. They discover themselves as being already there. The creation of empty places is their way of discovering their closure and of relating to the world. They are the re-entry of themselves into a world that would not exist if they did not occupy it. Spencer-Brown 1969; Leibniz 1714, § 15

5 Communication

18. Monads communicate with each other by each communicating with God they find within themselves. This is Leibniz's solution to the question of how the greatest possible order is possible with the greatest possible diversity. Leibniz 1714, § 58; Serres 1968

19. Without God we must double communication. Communication is the operation of any one monad and at the same time the medium of all other monads operating the same way. There is both micro-diversity and complexity of communication in any one communication, i.e. no operation without its medium, no step without recursion, no operation without repetition among similar operations. von Foerster 1971, 1976, 1980; Luhmann 1997a; White 1992, 4

20. Or else, there is no communication without meta-communication linking it to possible and potential communication while still indicating itself. Ruesch and Bateson 1951; Barel 1989

21. There is no information if the selection of a message is not evaluated against the set of possible messages, be it a technically given alphabet or a socially constructed context. Shannon 1948; Baecker 2013

22. Communication follows a pragmatic calculus whose axioms are obeyed in successful, and broken in disturbed, communication. Watzlawick, Beavin and Jackson 1967, 24

6 Negativity

23. The closure of monads produces the selection of a perspective on the world together with the excess of reference to everything that is excluded by this selection. Therefore, what is excluded is also included and becomes the set of possible contexts for interpreting and continuing the communication at hand. Luhmann 1990, 81; Luhmann 1999

24. The society of monads is the society of their distinction. Negativity (Aristotle's *stérēsis*), sustained by the co-presence of excess and lack, of correspondence and contradiction, separates and relates them. Aristotle, Physics, Book A, chaps. 5–9; Hegel 1813, Book 2, chap. 2, C

25. The calculus of monads, by re-entering the distinction into its form, is necessarily indeterminate such as to let an object of perception correspond with the non-formalised. Spencer-Brown 1969, 57

26. Communication oscillates between its selections and its contexts to both memorise and stay indeterminate. Spencer-Brown 1969; Luhmann 1997b

7 Medium and Form

27. Monads simultaneously produce their operation and the medium they operate in, so that a distinction between the two becomes possible. That distinction is another operation again producing its medium, which may or may not be the same. The tight coupling of the operation is complemented by the loose coupling of the medium, thus presenting the operation with its own alternatives. Heider 1959

28. The doubling of communication in terms of operation and medium is taken up, repeated and secured in all three dimensions of meaning: the temporal, the social, and the factual. Monads navigate between the reversible and the

irreversible present, between consensus and dissent with others, and between objective certainty and ecological uncertainty. Luhmann 1980

29. The operation of communication can no more determine the medium than the medium can determine the operation. Being doubled, the operation dissolves. At any given moment, communication is forcing its own decay. At any given moment, monads are left with their choices. Luhmann 1995, 290

30. There is a calculus of communication practised by monads which at any given moment looks at a form within the medium of its possibility:

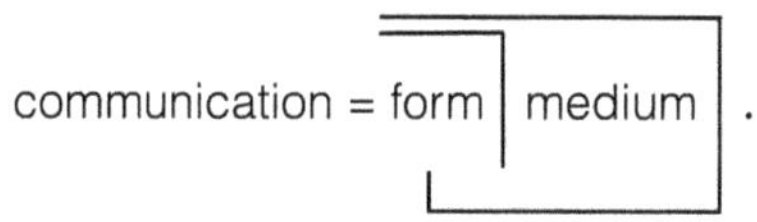

Communication proceeds by processing the difference between a form and its medium. The indication of a 'form'—which is itself an indication, with its distinction and the space it brings to the fore—reveals not only alternative possibilities of communication that did not exist before, but also its own preconditions that can be further explored. Heider 1959; Spencer-Brown 1969; Luhmann 2012, 88; Lehmann 2018; Baecker 2024

8 Oscillation

31. The monad is the oscillation between its form and its medium. It oscillates between its memory and the imaginary, if not the unknown. It will never know whether its content equals its intent. White 1995

References

[1] Aristotle. 2008. *Physics*. Translated by Robin Waterfield, edited by David Bostock. Oxford: Oxford University Press.

[2] Aristotle. 2002. *Metaphysics*. Translated by Joe Sachs. 2nd edition. Santa Fe, NM: Green Lion Press.

[3] Baecker, Dirk. 2013. "Systemic Theories of Communication." In *Handbooks of Communication Science, vol. 1: Theories and Models of Communication*, edited by Paul Cobley and Peter J. Schulz, 85–100. Berlin: de Gruyter Mouton.

[4] Baecker, Dirk. 2024. "Monadologie im Formkalkül." https://www.researchgate.net/publication/379873685_Monadologie_im_Formkalkul. Preprint.

[5] Barel, Yves. 1989. *Le paradoxe et le système: Essai sur le fantastique social.* 2nd enlarged ed. Grenoble: Presses Universitaires de Grenoble.

[6] Cheung, Tobias. 2009. "System, Mikrooperator und Transformation: Leibniz' gemeinsames Ordnungsdispositiv der Monade und des Lebendigen im naturgeschichtlichen Kontext." In *Der Monadenbegriff zwischen Spätrenaissance und Aufklärung*, edited by Hanns-Peter Neumann, 143–201. Berlin: de Gruyter.

[7] Cuntz, Michael. 2022. "Monadologie und Moderne: Bemerkungen zu Leibniz' und Tardes Entwürfen der bestmöglichen Welt im Zeitalter ihres Verschwindens." In *Formen des Ganzen*, edited by Eva Geulen and Claude Haas, 159–176. Göttingen: Wallstein

[8] Deleuze, Gilles. 1993. *The Fold: Leibniz and the Baroque.* Foreword and translation by Tom Conley. Minneapolis: University of Minnesota Press.

[9] Democritus. 2010. "Fragments." In Daniel W. Graham, *The Texts of Early Greek Philosophy: The Complete Fragments and Selected Testimonies of the Major Presocratics*, 516ff. 2 vols. Cambridge, UK: Cambridge University Press.

[10] Epicurus. 1853. "Letter to Herodotus." In Diogenes Laertius, *The Lives of the Philosophers*, trans. C. D. Yonge, Book 10, Sections 35–83. London: Henry G. Bohn.

[11] Frege, Gottlob. 1891. "Funktion und Begriff." In *Funktion—Begriff—Bedeutung*, edited by Max Textor, 2–22. 2., durchges, Aufl. Göttingen: Vandenhoeck & Ruprecht, 2007.

[12] Hecht, Hartmut. 1998. "Anmerkungen und Nachwort." In Gottfried Wilhelm Leibniz, *Monadologie*, 66–120. Translated by H. Hecht. Stuttgart: Reclam.

[13] Hegel, Georg W. F. 2010. *The Science of Logic.* Translated and edited by George di Giovanni. Book II. Cambridge: Cambridge University Press. First published 1813.

[14] Heider, Fritz. 1959. "Thing and Medium." *Psychological Issues* 1 (3): 1–34.

[15] Lehmann, Maren. 2018. "Eine schicksalhafte Verbindung: Mathematik und Soziologie." In *Mathematik und Gesellschaft: Historische, Philosophische und Didaktische Perspektiven*, edited by Gregor Nickel et al., 81–92. Wiesbaden: VS Springer.

[16] Leibniz, Gottfried Wilhelm. 1714. *Monadology: A New Translation and Guide.* Translated by Lloyd Strickland. Edinburgh: Edinburgh University Press, 2014.

[17] Leucippus. "Fragments." In Daniel W. Graham, *The Texts of Early Greek Philosophy: The Complete Fragments and Selected Testimonies of the Major Presocratics*, 516ff. 2 vols. Cambridge, UK: Cambridge University Press, 2010.

[18] Lucretius. *De Rerum Natura.* Translated by William E. Leonard. New York: E. P. Dutton, 1916.

[19] Luhmann, Niklas. 1980. "Temporalstrukturen des Handlungssystems—Zum Zusammenhang von Handlungs- und Systemtheorie." In *Verhalten, Handeln und System: Talcott Parsons' Beitrag zur Entwicklung der Sozialwissenschaften*, edited by Wolfgang Schluchter, 32–67. Frankfurt am Main: Suhrkamp.

[20] Luhmann, Niklas. 1990. *Die Wissenschaft der Gesellschaft.* Frankfurt am Main: Suhrkamp.

[21] Luhmann, Niklas. 1995. *Social Systems.* Translated by John Bednarz. Stanford, CA: Stanford University Press.

[22] Luhmann, Niklas. 1997a. "Selbstorganisation und Mikrodiversität: Zur Wissenssoziologie des neuzeitlichen Individualismus." *Soziale Systeme: Zeitschrift für Soziologische Theorie* 3 (1): 23–32.

[23] Luhmann, Niklas. 1997b. "The Control of Intransparency." Translated by M. P. van der Marel and A. Zÿlstra. *Systems Research and Behavioral Science* 14 (6): 359–371.

[24] Luhmann, Niklas. 1999. "The Paradox of Form." In *Problems of Form*, edited by Dirk Baecker, 15–26. Stanford, CA: Stanford University Press.

[25] Luhmann, Niklas. 2012. *Theory of Society.* Translated by Rhodes Barrett. Vol. 1. Stanford, CA: Stanford University Press.

[26] Mahnke, Dietrich. 1937. *Unendliche Sphäre und Allmittelpunkt: Beiträge zur Genealogie der mathematischen Mystik.* Halle/Saale: Niemeyer.

[27] Novalis. 1795/96. "Philosophische Studien 1795/96 (Fichte-Studien)." In *Werke, Tagebücher und Briefe Friedrich von Hardenbergs, vol. 2, Das philosophisch-theoretische Werk*, edited by Hans-Joachim Mähl, 7–209. Munich: Hanser, 1978.

[28] Pearce, W. Barnett, and Vernon E. Cronen. 1980. *Communication, Action, and Meaning: The Creation of Social Realities.* New York: Praeger.

[29] Plato. *Sophist.* Translation, introduction and notes by Nicholas P. White. Indianapolis, IN: Hackett, 1993.

[30] Plato. *Philebus.* In *Great Dialogues of Plato,* translated by W. H. D. Rouse. London: Penguin, 1999.

[31] Pythagoras. "Fragments." In Daniel W. Graham, *The Texts of Early Greek Philosophy: The Complete Fragments and Selected Testimonies of the Major Presocratics*, 905ff. 2 vols. Cambridge, UK: Cambridge University Press, 2010.

[32] Ruesch, Jurgen, and Gregory Bateson. 1951. *Communication: The Social Matrix of Psychiatry.* New York: Norton.

[33] Scheier, Claus-Artur. 2016. *Luhmanns Schatten: Zur Funktion der Philosophie in der medialen Moderne.* Hamburg: Meiner.

[34] Schuller, Marianne, and Gunnar Schmidt. 2003. *Mikrologien: Literarische und philosophische Figuren des Kleinen.* Bielefeld: transcript.

[35] Serres, Michel. 1968. "La communication substantielle démontrée more mathematico." In *Hermès I: La Communication*, 154–164. Paris: Minuit.

[36] Shannon, Claude E. 1948. "A Mathematical Theory of Communication." *Bell System Technical Journal* 27: 379–423 and 623–656.

[37] Spencer-Brown, George. 2008. *Laws of Form.* Leipzig: Bohmeier Verlag. First published 1969 by George Allen & Unwin.

[38] Varga von Kibéd, Matthias, and Rudolf Matzka. 1993. "Motive und Grundgedanken der 'Gesetze der Form.'" In *Kalkül der Form*, edited by Dirk Baecker, 58–85. Frankfurt am Main: Suhrkamp.

[39] von Foerster, Heinz. 1971. "Computing in the Semantic Domain." *Annals of the New York Academy of Sciences* 184: 239–241.

[40] von Foerster, Heinz. 1976. "Objects: Tokens for (Eigen-)Behaviors." *Cybernetics Forum* 8: 91–96.

[41] von Foerster, Heinz. 1980. "Epistemology of Communication." In *The Myths of Information, Technology and Post-Industrial Culture*, edited by Kathleen Woodward, 18–27. Madison, WI: Coda.

[42] Watzlawick, Paul, Janet H. Beavin, and Don D. Jackson. 1967. *Pragmatics of Human Communication: A Study of Interactional Patterns, Pathologies, and Paradoxes.* New York: Norton.

[43] Weiner, Sebastian F. 2016. *Aristoteles' Bestimmung der Substanz als logos.* Hamburg: Meiner.

[44] White, Harrison C. 1992. *Identity and Control: A Structural Theory of Action.* Princeton, NJ: Princeton University Press.

[45] White, Harrison C. 1995. "Network Switchings and Bayesian Forks: Reconstructing the Social and Behavioral Sciences." *Social Research* 62 (4): 1035–1063.

[46] Whitehead, Alfred North. 1967. *Science and the Modern World: Lowell Lectures, 1925.* New York: Free Press.

[47] Wieland, Wolfgang. 1962.*Die aristotelische Physik: Untersuchungen über die Grundlegung der Naturwissenschaft und die sprachlichen Bedingungen der Prinzipienforschung bei Aristoteles.* 3rd ed. Göttingen: Vandenhoeck & Ruprecht, 1992.

The Numeric Forms of James Algebra

William Bricken
Unary Computers
Snohomish Washington USA
william@iconicmath.com

James algebra is an experiment in constructing a complete algebra of numbers, including integer, fractional, real and imaginary expressions, using as closely as possible the notation and structural principles of *Laws of Form*. Distinctions are conceptualized as two types of containers. *Containment* is the only structural concept. Void-based forms and pattern-driven substitution are maintained while the reduction provided by the primary arithmetic is suppressed. Three pattern axioms are sufficient and no complex theorems are needed. Conventional numeric concepts are necessary only when transcribing from the container calculus to numeric algebra. There are many surprises, including the use of a non-numeric unit, a single generic inverse, base-free exponents, and the fundamental role of a forgotten *additive* imaginary constant.

Notation...is extremely important in mathematics.
A seemingly modest change of notation may suggest
a radical shift in viewpoint.[1]
— Barry Mazur (2003)

Jeffrey James and I developed James algebra as his 1993 Master's Thesis at University of Washington, *A Calculus of Number Based on Spatial Forms*. The exotic forms of this **boundary algebra** are described in depth in a three volume series *Iconic Arithmetic*.[2] Much of this article comes from these books. James numbers take their inspiration from *Laws of Form*, the seminal work of George Spencer Brown, and from the work of Professor Louis Kauffman at the University of Illinois at Chicago. James algebra is a radical reconceptualization of how we represent and think about conventional numeric operations `{+, −, x, ÷, ^, √, log}`. James forms are added together simply by putting them together into the same container. Multiplication is represented by a specific configuration of boundaries. Pattern matching of containment relations and substitution of equals is the only method of transformation.

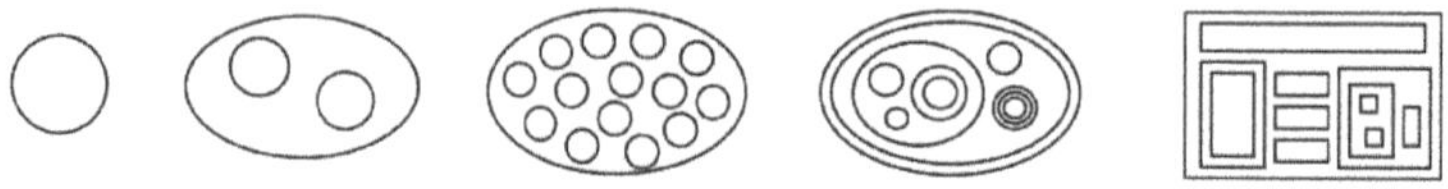

some patterns of containment

The axioms of modern algebra evolved in tandem with linear typography and with sequential (implicational) thinking. The axiomatic structure of algebra and set theory do not capture the *essence* of what numbers are nor how they work since neither system rests upon the Additive Principle that a sum is the collection of its parts. Our conventional perspective on arithmetic, the one currently taught in schools, is extremely useful for business and for scientific professionals, quite useful for sequential computers, and definitely a nuisance for educators, for young students, and for parallel, concurrent and distributed computation.

After a short introduction, I'll describe the types of containment patterns, their structural features, and the James axioms followed by a few computational examples. The article will end with a short comparison of James algebra to Spencer Brown's axioms and show some of the multidimensional dialects of boundary math. If you prefer to see the rules of the game before you meet the players, have a look at Figure 4.

1. Postsymbolic Math

Euler was first to propose a diagrammatic method of logic. His **Euler diagrams** associate embedded and overlapping circles with logic syllogisms. He explains:

> The foundation of all these forms is reduced to two principles, respecting the nature of *containing* and *contained*. I. Whatever is in the thing contained must likewise be in the thing containing, and II. Whatever is out of the containing must likewise be out of the contained.[3]

In the nineteenth century, the trauma of non-Euclidean geometries shook the mathematical world so fundamentally that the ancient Greek perspective of deriving mathematical knowledge from diagrams and manipulatives was completely abandoned, in favor of purely symbolic approaches. Nobel Laureate Herbert Simon: "Rigor, it was believed, called for reasoning to be formalized in symbols arranged in sentences and equations." [4] Hilbert's program at the turn of the twentieth century set out to express mathematical reasoning in finite strings of symbols. Mathematical diagrams and other sensory and experiential forms were widely purged from rigorous mathematics.

The **syntax/semantics barrier** is deeply implicated in the migration to linear structure. The meaning of words and strings of symbols became entirely separate from the words and symbols themselves. Symbols require augmentation, meaning must be added separately. Here's a perspective from Larkin and Simon's famous paper, *Why a diagram is (sometimes) worth ten thousand words*:

> The fundamental difference between our diagrammatic and sentential representations is that the diagrammatic representation preserves exactly the information about the topological and geometric relations among the components of the problem, while the sentential representation does not.[5]

Strings of symbols do impact meaning by limiting how we think about what we are describing. Logicians Barwise and Etchemendy:

> Diagrams are physical situations.... the user can simply read off the facts from the diagram as needed. This situation is in stark contrast to sentential inference, where even the most trivial consequence needs to be inferred explicitly.[6]

The forms of James algebra are **containers** that support the visual relation of inside and outside. *James forms are spatial.* They are iconic rather than symbolic. The typographic representation used herein shows containers as string delimiters such as ( ), with the image of the container broken into a left and a right half. However, the sequencing and the fracturing of boundaries is an *accident* of the way that our typewriters and our textual languages are constructed, and a potentially confusing distortion of the image of a container.

2. James Forms

Our theme is that numbers are sensory, diagrammatic, experiential. In James algebra **forms** are configurations of container boundaries. All the possible containment patterns constitute the **language** of James forms. Structurally different forms of containment are defined by pattern equations. The structure of a form can be changed only by following specified pattern rules. **Axioms** are transformation rules that are asserted as design decisions. **Theorems** are convenient transformations that follow directly from the axioms. **Frames** are containment structures that provide a conceptual organization of types of forms. **Interpretations** are ways to convert James forms into conventional expressions within the arithmetic and algebra of numbers. Some James forms can be interpreted as imaginary numbers, some can be interpreted as being non-numeric.

First and foremost, James forms are **void-based**. **Void-equivalent** structures and patterns are inactive and irrelevant to meaning. These forms exist solely in notation. Void-equivalent forms can be considered to be *illusions* that arise from an empty page.

James forms are *patterns* of containment that can be represented by *physical* containers. A container is an *object* from the outside and a containment *process* from the inside. Concepts are *networks* of contains relations. Every container is *independent* but for its containment relations with other containers.

A **calculus** consists of a notation for representing objects, a collection of permitted transformations and a collection of basic facts. Container objects are delimited by their boundaries, while basic facts are empty containers. From the perspective of James calculus, common numbers and their arithmetic are fully described by putting containers into other containers and simplifying the resulting containment relations by following the pattern rules defined by the three James structural axioms. In James algebra, there are no instances of counting, ordering or grouping because there is no imposition of structure other than that of containment. Importantly, only one axiom permits rearrangement of structure, the forte of string languages. The other two axioms as well as two convenience theorems are void-based, they eliminate structure by erasure/deletion, by casting meaningless structure into *void*.

boundary	*unit*	*interpretation*	*operator*	*interpretation*
round	( ), o	1	(A)	$\#^{A}$
square	[ ]	$-\infty$	[A]	$\log_{\#} A$
J	J	$\log_{\#} -1$	(J [A])	-A

Figure 1: *James units and operations*

3. Container Types

James algebra incorporates two types of boundaries, represented here by two types of typographic delimiters. The **round boundary**, ( ), can be interpreted as the archetypal numeric unit, One. The **square boundary**, [ ], is an archetypal non-numeric unit. The names and shape of these boundaries is for typographic convenience and in no way suggests features of geometry. Boundary forms are icons. Pictorial forms trigger not only different conceptual models, they trigger different physiological processes. Transcription is therefore more than a cognitive shift, it implicates different perceptual systems and a different behavioral vocabulary.

Figure 1 shows two forms of the **empty round boundary** and its interpretation. When the round boundary holds contents, it is an operator that can be interpreted as exponentiation of an arbitrary base. The arbitrary base is represented by a hash mark, #. With no content a container is a constant. Within the language of boundary forms, we would say that the whole arises out of framing nothing.

Natural numbers are ensembles of round units. Round boundaries (alternatively round containers, round brackets) *accumulate*. The emptiness inside a round unit is intended to look like nothing, i.e. the absence of quantity that is conventionally represented by 0. Without interpretation, James forms are simply *patterns of containment*. James forms map one-to-many onto conventional numeric expressions, with interpretation a choice of the reader. The empty round boundary, for example, has two interpretations, as a unit and as an exponential operator on nothing. James algebra makes no distinction between the object *one* and the function *raise-to-the-zero-power*. The concepts of objects and operators are replaced by the containment relation.

( ) ☞ 1   ( ) ☞ $\#^{0} = 1$

The **empty square boundary** is not numeric. If it were to be interpreted it would be something like *negative infinity*. The axioms are designed so that a lone square boundary does not normally occur. When the square boundary has contents, it can be interpreted as a logarithm with an arbitrary base. The third row in Figure 1 is the constant J that will be described later as a generic inverse. It can be interpreted directly as the logarithm of negative one, $\log_{\#} -1$, again with an arbitrary base.

Base-free Exponents

The operation of *raising-to-a-power* can be **base-free** just as an algebraic variable can be number-free when it applies to all numbers. We have interpreted the round boundary as raising # to the power of its contents

(A) ☞ $\#^{A}$

For consistency, the square boundary then is interpreted as the base-free logarithm of its contents.

[A] ☞ $\log_{\#} A$

The *hash mark*, #, is a meta-symbol that is neither part of the James notation nor part of the interpretation. In our interpretation, boundaries can stand in place of exponential and logarithmic expressions. In contrast, transformation within James forms is independent of the concept of a base. In effect, # is a convenience that permits the *interpretation* to remain uncommitted to a value of a base. The concept of base itself is an *interpretative add-on*, not part of the container algebra. To *interpret* a round boundary as a conventional expression, we need to provide both the operation (raise-to-a-power) and the base (raise-*this*-to-a-power).

The hash-mark is a special symbol for the form (( )). The value of a **double round boundary** is the value of the base assigned to a single round boundary.

( ) ☞ $\#^{0} = 1$

(( )) ☞ $\#^{1} = \#$

The concept is strange and bears repeating. Within the exponential interpretation, (( )) is not a particular number, rather it is a specific form that we can safely impose any numeric value upon, so long as we apply that value consistently during interpretation. The hash-mark is not a variable, not an arbitrary number; it is a meta-token that supports an *arbitrary interpretation*.

Thus the *numeric* value of $\#^A$ is indeterminate until # is fixed by assignment to a concrete number. Within the James algebra, in contrast, we are not raising-to-the-power-of A, we are simply putting A into a round container.

There is a conventional way to understand the implications of an arbitrary base. Any choice of base is just a constant factor away from any other choice of base. The constant of proportionality, $1/\log_B N$, relates the chosen base B to an arbitrary base N.

$$\log_N A = (1/\log_B N) \times \log_B A$$

The constant of proportionality is simply a scaling factor, the same type of scaling factor that we use to convert between units of measurement, such as measuring in kilometers rather than in miles. Educator Paul Lockhart:

> Choosing a base for our exponents is exactly the same as choosing a measuring unit — essentially we are measuring the size of a number by how many decimal digits it has.... If what you want is to convert multiplication to addition, then any base is as good as any other.[7]

Similarly, the size of a unit is arbitrary, a perspective supported by thinking of units as a measure of distance between 0 and what we call "1". In a void-based system, however, the "size of" ( ) is the "distance" between non-existence and existence.

Void-equivalence

The word *void* is in our shared descriptive metalanguage, however it is not part of the James system. **Metalanguage** talks *about* a formal system but the concepts and expressions within a metalanguage are not *within* the system itself. We cannot refer to *void* within the James algebra because there is no form with which to associate a reference. Within the algebra, an empty container itself provides indirect reference to its non-existent contents. The arithmetic rule of Void Inversion provides an example of a void-equivalent form.

([ ]) = *void*

In all cases, equations such as Void Inversion are more properly written without recourse to the metalanguage.

([ ]) =

Equality

If we begin with nothing, where does the *equal sign* = come from? Equality is the absence of distinction. The base case is

$$void = void$$

which can be more elegantly stated without the crutch of metalanguage as

$$=$$

Equality constructs two indistinguishable perspectives. Since *void* cannot be replicated, the two perspectives on either side of the equal mark are created by the mark itself. Equality is the distinction that does not distinguish inside from outside. *Equality is a distinction that does not distinguish.*

Inequality identifies a change in viewing perspective: looking from the outside vs looking from the inside. More abstractly, the boundary definition of the ≠ sign is that it identifies a **difference.** ≠ declares that crossing a boundary (whether physically or perceptually) makes a difference. Difference, in turn, is a fundamental property of distinction.

The idea of constructing a distinction is visceral. We see an *object* from the outside. When we cross the boundary to the inside, we can no longer see the object. Instead we see an *environment*. In an empty environment we see nothing. **Existence** is created (or observed if you prefer) by the act of crossing a distinction from inside to outside.

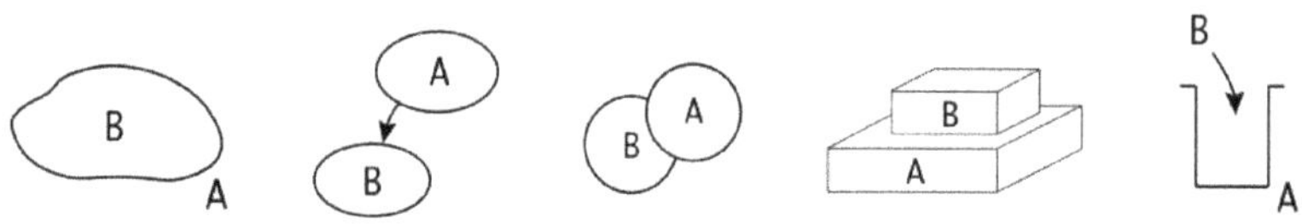

Containment

The only relation within the James calculus is that of **containment**, a *minimal conceptual basis* consisting of one binary relation `A contains B`. The `contains` relation is quite general. When expressed within logic, containment can be interpreted as `implies`. When expressed as an acyclic network, containment is `directly-connected-to`. When expressed as a set, it's called `is-a-member`. When expressed as a number, it is `successor`. When expressed as a map, it's `shares-a-common-border`. Within the context of a pile of blocks, `contains` becomes `supported-by`. When described as an abstract mathematical structure, a form is an *unordered rooted tree*. These interpretations share a collection of common characteristics that are concretized by the properties of physical containers. The fundamental *concept* underlying containment is **distinction**: a container distinguishes inside from outside.

Accumulation

Natural numbers are grounded by the concept of distinction. The concept of *repetition* generates the plethora of distinctions. The construction of whole numbers proceeds in the same manner as it has for millennia, via the accumulation of units. Like all tally systems, an essential property of James numbers is that forms add simply by being placed together. Formal grounding for the Additive Principle is provided by the Accumulation axiom of James arithmetic:

$$()() \neq ()$$

A consequence of accumulation is that all James forms that can be interpreted as numbers will include o in the deepest level of their canonical form.

We can call the multiplicity of forms grouped together by a common boundary the **cardinality** of that boundary. The concept of cardinality applies to any container. Cardinality requires only that every container is independent but for its containment relations. Accumulation does not depend upon labeling each unit. Cardinal equality is defined by matching structural patterns. Equivalence does not depend upon counting.

Unification

As a type of unit, [ ] must have different additive properties than ( ), otherwise the two would be indistinguishable. Round-brackets provide a tally via accumulation. In contrast, [ ] *dominates* rather than accumulates. We can express this by an iconic equation, one that reduces apparent multiplicity to unity. The Unification axiom defines the behavior of square-brackets.

$$[\,]\ [\,] \Rightarrow [\,]$$

More generally $$[\,]\ \ A \Rightarrow [\,]$$

Although this axiom is an equality, it is only used in the direction of simplification. The square-bracket does not support replication. It is a semantic error to believe that more than a singular [ ] can exist, given its pervasive domination of form. Square-brackets are blind to multiplicity, making them clearly non-numeric. Strictly speaking, [ ] does not *absorb* the forms within its environment since no forms within the same container interact. More accurately, [ ] completely fills its container, so that the outer container can support no other contents. From the perspective of the outer container, Unification asserts that there is only one empty square-bracket whereas Accumulation asserts that every round-bracket makes a difference.

Unification provides an interpretation of conventionally extreme numeric cases, particularly when 0 is implicated. Our current interpretation reads square-brackets as taking a logarithm. An *empty* square-bracket represents the logarithm of zero, $\log_{\#} 0$. We can avoid the difficulty of mixing numeric and non-numeric forms by the simple expedient of leaving [ ] *undefined for interpretation*. However, [ ] is a valid form to manipulate within James algebra. A subtle constraint has been introduced by the interpretation of [ ]: *# is greater than* 1. When # is between 0 and 1, say ½, the fractional base changes the sign of the result.

$$\log_{\#} 0 = -\infty \qquad\qquad \log_{1/\#} 0 = +\infty$$

For now we'll assume that # > 1 when # is the base of a logarithm.

Interpretation

Figure 2 provides a quick introduction to the **interpretation** of boundary configurations that we will use. These patterns unify counting, adding, multiplying, raising to a power, and the assortment of inverse operations under the three James axioms. The J-frame (J [...]) is a generic inverse that can be interpreted as a negation, division or logarithmic operator depending upon what it contains. A convenient abbreviation for (J [A]) is <A>. Figure 2 includes this abbreviation for ease of reading.

4. Frames

Frames are the structural framework underlying the James axioms and, indeed, underlying most stable James forms. Frames are forms for which an outer round-bracket contains at least two content forms, at least one of which is enclosed in a square-bracket, for example (A [B]). Otherwise, a form is either simply nested or accumulating. The **generic structure** of a frame is

(*frame-type* [*frame-content*])

expression	☞ *form*	*abbreviation*
	UNITS	
1	( )	
$-\infty$	[ ]	
-1	(J)	<( )>
$\log_{\#} -1$	J	
	INVERSE	
A	A	
$-A$	(J [A])	<A>
1 / A	((J [[A]]))	(<[A]>)
	ARITHMETIC	
A + B	A B	
A – B	A (J [B])	A <B>
A x B	([A] [B])	
A / B	([A](J [[B]]))	([A]<[B]>)
	BASE	
B^A	(([A] [[B]]))	
B^{-A}	((J [A] [[B]]))	(([[B]] [<A>]))
$B^{1/A}$	(((J [[A]]) [[B]]))	(([[B]] <[A]>))
$\log_B A$	([[A]] (J [[[B]]]))	([[A]]<[[B]]>)
	EMBEDDED BASE	
#	(( ))	
$\#^A$	(A)	
$\log_{\#} A$	[A]	

Figure 2: *Algebraic operations as patterns of containment*

Frames provide a conceptual scaffolding as well as a structural template. The axioms of James algebra (displayed in Figure 4) specifically modify the structural relations of the round and square frame boundaries together. This permits the base-free interpretation of exponents and logarithms. When boundary configurations are interpreted as numeric operations, the conventional axioms of algebra can be organized as varieties of frames.

FRAMES		
([ ]) = *void*	**void frame**	*axiom*
([A]) = A	**inversion frame**	*axiom*
(A [B C]) = (A [B]) (A [C])	**arrangement frame**	*axiom*
(A [ ]) = *void*	**dominion frame**	*theorem*
([A][o]) = A	**indication frame**	*theorem*
([A][B]) = $A .._{B}.. A$	**cardinality frame**	*theorem*
(o [B])	**unit magnitude frame**	*magnitude*
(A [B])	**magnitude frame**	*magnitude*
((J)[B])	**decimal frame**	*magnitude*

Figure 3: *Types of frames*

Frame Types

Figure 3 identifies several types of frames. Since our interpretation takes round-brackets to be powers and square-brackets to be logarithms, a generic frame represents an exponential expression, $B\#^{A}$.

(A [B]) ☞ $\#^{A+\log_{\#}B} = \#^{A} \times B$

([(A)][B]) ☞ $\#^{A} \times B$

The *cardinality* frame expresses multiplication in general by dispersing replicas of specific forms.

([*form*][*cardinality*]) ☞ *form* x *cardinality*

([A][N]) = $A .._{N}.. A$

The *dominion* frame is the base-case of cardinality. The *indication* frame plays a central role in counting.

([A][ ])	☞	A x 0 = 0	*cardinality* 0
([A][o])	☞	A x 1 = A	*cardinality* A
([A][B])	☞	A x B	*cardinality* AxB

The *magnitude* frame expresses the exponential structure of polynomial digits. The magnitude frame thus resembles scientific notation, and disperses replicas of exponential forms.

Axioms

ARITHMETIC

`() () ≠ ()`	**unit accumulation**
`([]) = [()] =` *void*	**void inversion**
`(J) () =` *void*	**unit reflection**

ALGEBRA

`([A]) = [(A)] = A`	**inversion** *enfold ⇄ clarify*
`(A [B C]) = (A [B]) (A [C])`	**arrangement** *collect ⇄ disperse*
`A (J [A]) =` *void*	**reflection** *create ⇄ cancel*

Theorems

`(A [ ]) =` *void*	**dominion** *emit ⇄ absorb*
`(J J [A]) = A`	**involution** *wrap ⇄ unwrap*

Figure 4: *James axioms and theorems*

(*power* [*value*]) ☞ *value* x $base^{power}$

`( A [N]) = (A)..`$_{N}$`..(A)`

Unitary magnitude multiplies a given magnitude A by the elected base #.

`( o [A])` ☞ $\#^{1+\log_{\#}A} = \# \times A$

5. Pattern Axioms

Figure 4 shows the pattern axioms of James algebra, the transformations that are designed to map onto the behavior of numeric arithmetic and algebra. The three boundary arithmetic axioms define relationships between empty containers. Unit accumulation reiterates that units do not condense. Void inversion asserts that the two types of containers are inverses. Unit reflection identifies the round unit inverse. It can be easily understood by interpreting (J) as –1.

In boundary algebra two of the three axioms define the interplay between round and square boundaries, while the third defines the behavior of J as a generic inverse. It is convenient to be able to identify the direction of application of each axiom when describing computational steps. Figure 4 also provides these names. All transformations of containment patterns are essentially simple. There are no particularly subtle or creative theorems. The two theorems in Figure 4 are convenience lemmas. The most important characteristic of these axioms is that Inversion and Reflection specify how to *delete* structure. Both implicate only one form (labeled A), so that they both require only simple pattern-matching. Remarkably, this leaves all of the complexity of numeric algebra isolated in one pattern transformation, Arrangement.

From the structural perspective, containers have no inherent meaning other than their ability to contain. Meaning is off-loaded onto an interpretation, so that we may read containment structures as physical forms, for example as collections of nested boxes. The interpretation is dragged along with the valid transformations. Equality is defined by *permitted structural transformations*. This is in contrast to the usual interpretation of the equal sign, = , as invariance of numeric value.

In general, a **variable** within an algebra stands in place of an *arbitrary* form. In James algebra, this idea is slightly more complex. Since the algebra is void-based, a variable might stand in place of nothing. Variables that are not void-equivalent stand in place of a single container and its contents. It is a violation of the independence of containment to have multiple forms standing in the same space without an outermost container. However, it is often typographically convenient to leave the outermost container implicit. We can display, for example, either ([oooo]) or oooo, with the understanding that the outermost container of oooo may be unwritten but it is certainly present.

Inversion

The inverse of the round boundary ( ) is the square boundary [ ]. The round/square pair of boundaries is jointly called an **inversion pair**. Inversion pairs are co-defined to mutually annihilate, regardless of which contains which. This containment relation is expressed by the Inversion axiom:

([A]) = [(A)] = A

Inversion pairs are void-equivalent and can be freely created and deleted. Void Inversion is a special case of Inversion for which A = *void*.

Inversion pairs provide an assured outermost boundary for any generic collection of boundary forms, making all forms conceptually singular.

A B C = ([A B C]) = [(A B C)]

Round/square frames that do not reduce, such as (A [B]), are a **central organizing concept** in James algebra.

Inversion pairs can be interpreted as any pair of conventional inverse functions, for example, ( ) as double and [ ] as half.

([ ]) ☞ double(half(zero)) = zero

([4]) ☞ double(half(four)) = four

[(A)] ☞ half(double(A)) = A

We can also choose to engage in a non-numeric interpretation such as ( ) standing in place of "open the door" and [ ] representing "close the door". Each unit boundary defines a new state of the door, while void inversion returns the door to its original state.

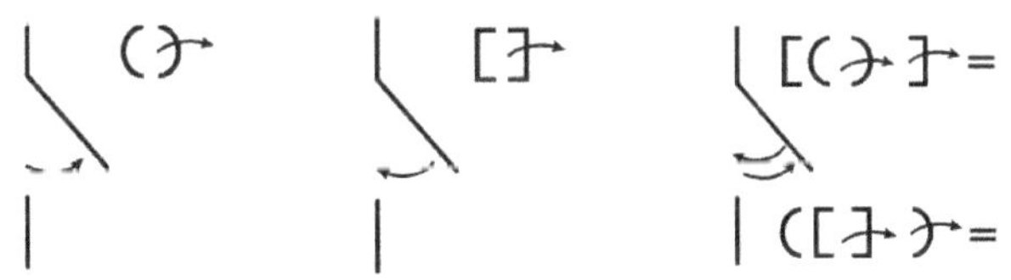

The separate contents of round-brackets can be seen to represent multiplications. For example,

(A B C) ☞ $\#^{A+B+C} = \#^{A} \times \#^{B} \times \#^{C}$

The form of multiplication is more clearly illustrated when the variables are themselves within a square-bracket.

([A][B][C]) ☞ $\#^{\log_{\#} A + \log_{\#} B + \log_{\#} C}$

$= \#^{\log_{\#} A} \times \#^{\log_{\#} B} \times \#^{\log_{\#} C} = A \times B \times C$

Every form can be interpreted as a multiplication. In the case of adding three forms, each of A, B, and C can be enclosed by an application of enfold, and if necessary the entire form can also be `enfolded`. Here we convert addition into multiplication by applying `enfold` four times:

```
A B C = [( [(A)] [(B)] [(C)] )]
        ☞  A+B+C = log# (#^A x #^B x #^C)
```

$$A\ B\ C = [(\ [(A)]\ [(B)]\ [(C)]\ )] \quad ☞ \quad A+B+C = \log_{\#} (\#^A \times \#^B \times \#^C)$$

When square-brackets are interpreted as $\log_{\#}$, we might say that `( )` is 1, and that the logarithm of 1 in any base is 0.

$$[(\ )] = \textit{void} \quad ☞ \quad \log_{\#} 1 = 0$$

$$([\ ]) = \textit{void} \quad ☞ \quad \#^{\log_{\#} 0} = 0$$

In the general case,

$$[(A)] = A \quad ☞ \quad \log_{\#} \#^A = A \times \log_{\#} \# = A \times 1 = A$$

$$([A]) = A \quad ☞ \quad \#^{\log_{\#} A} = A$$

The perspective we are taking is that *exponentials and sums are equally fundamental*, while multiplication is a composite and therefore derivative operation.

Arrangement

To complete an operational description of round- and square-brackets, we will need an axiom that specifies how the two boundaries interact in configurations more complex than the double nesting of Inversion. The James axiom of Arrangement specifies an invariant structure across frames.

$$(A\ [B\ C]) = (A\ [B])\ (A\ [C])$$

The simplifying direction (from right to left) is `collect`, and the expanding direction (from left to right) is `disperse`.

The Arrangement axiom gives permission to take multiple forms out of the interior of a frame and put them into separate frames, so long as we carry their context, i.e. their frame type, `(A [...])`, along with them. Arrangement is roughly equivalent to conventional distribution, with the difference that the conventional distributive rule distributes multiplication over addition, while our interpretation of Arrangement distributes addition of logarithms over exponents. Due to the awkward conventional notation for exponents and logarithms, the interpretation of Arrangement in standard notation (plus the base-free #) is not pretty:

$$(A\ [B\ C]) = (A\ [B])\ (A\ [C]) \quad ☞$$

$$\#^{(A + \log_{\#}(B+C))} = \#^{(A+\log_{\#} B)} + \#^{(A+\log_{\#} C)} \quad \Rightarrow$$

$$\#^{A} \times (B + C) = (\#^{A} \times B) + (\#^{A} \times C)$$

Expressing the "A" form as [A] simplifies the transcription to look like the rule of Distribution:

$$([A][B\ C]) = ([A][B])\ ([A][C]) \quad ☞$$

$$A \times (B + C) = (A \times B) + (A \times C)$$

The pattern of Arrangement provides the **only structural complexity** within the James form. It is the only pattern that permits *rearrangement* as opposed to deletion/creation. It is the only pattern that differentiates the content of James frames. And it is the only pattern that introduces replication (of the form A) without changing cardinality.

Although an equality, the different transformation directions of Arrangement have very different objectives. `Collect` gathers forms that share the same type of frame, in effect making the frame-type a global property of the content forms. `Disperse` separates the frame-content so that individual forms can be transformed separately within individual frames.

Arrangement is much broader than conventional distribution since container forms support multiple conventional interpretations. We could express the distributive rule in a generic rather than a multiplicative form:

$$([A][B\ C]) = ([A][B])\ ([A][C]) \qquad \textit{multiplicative}$$

$$(\ A\ [B\ C]) = (\ A\ [B])\ (\ A\ [C]) \qquad \textit{generic}$$

This is not a significant structural difference, since any form can be bounded by [ ] via the Inversion axiom: A = [(A)]. There is a form of Arrangement that looks like conventional factoring, so we could call it the factored interpretation.

$$(A\ [(B)(C)]) = (A\ B)\ (A\ C)$$

This form is decidedly not factoring, the resemblance is purely superficial. Specifically, [(B)(C)] is not a multiplication, it is a logarithm.

$$[(B)(C)] \quad ☞ \quad \log_{\#} (\#^{B} + \#^{C})$$

Log does not distribute over addition, so this conventional form is as simple as it gets.[8] The factored version of Arrangement can be cast into the form of multiplication, yielding a conventional but unusual rule of exponents.

```
([(A)][(B)(C)]) = (A B) (A C)
```

☞ $\#^{A} \times (\#^{B} + \#^{C}) = \#^{A+B} + \#^{A+C}$

To make this interpretation more familiar to the eye, we can partition it over two conventional rules: distribution followed by the product rule of exponents.

$$N^{A} (N^{B} + N^{C}) = N^{A} N^{B} + N^{A} N^{C} = N^{A+B} + N^{A+C}$$

Reflection

The third axiom, Reflection, introduces the constant unit J, which can be interpreted as a *generalized inverse*. We have built addition, multiplication, and exponentiation out of the same two boundary containers. The inverses of these operations, subtraction, division and logarithms, are conventionally taken to be separate operations. However since the constructive operations are defined by containment relations, the same containment relations can be used to define the reductive operations. What is needed is an indicator of the direction of application. J is that indicator. J is not present in the constructive forms. The presence of J indicates that a constructive operation is reversed. This generic modification applies to all constructive operations because these operations are interpretations of boundary configurations rather than independent procedures within the boundary forms. The reductive operations themselves are also within the interpretation rather than within the boundary configuration.

Reversal of a functional interpretation is achieved by introducing a J-frame, (J [...]). The inversion itself does not apply to an entire form, rather it applies to that subform that is associated with the reduction.

add B:	`A B`
subtract B:	`A (J [B] )`
multiply by B:	`([A] [B] )`
divide by B:	`([A] (J [[B]] ) )`
exponent base-B:	`(( [A] [[B]] ))`
logarithm base-B:	`([[A]] (J [[[B]]]) )`

The J-frame is closely associated with the inversion pair that forms the J-frame. It is the presence of J within the inversion pair `(  [...])` that constructs the numeric inverse, otherwise the inversion pair is void-equivalent:

```
inversion:      ( [A])  =  A
J-inversion:    (J [A]) ☞ −A
```

J-inversion does not necessarily construct a numeric negative, so a transcription is necessary to get to −A. The minus sign that identifies negative numbers is but one instance of a more general concept. Conventional notation also makes the leap of associating negation with division in the expression $A^{-1} = 1/A$.

Reflected Forms

Figure 2 provides a visual overview of the structure of the reflection pairs for the common operations of arithmetic. For **add/subtract**, the Reflection axiom provides the operational cancellation. This axiom in fact *defines* J.

```
A − A = A + −A = 0   ☞   A (J [A]) = void
```

The product of A and its reciprocal 1/A reduces to the multiplicative identity 1. Here is the form of a reciprocal:

```
1/A ☞ ([o] (J [[A]])) = ((J [[A]]))              clarify

  A x 1/A ☞ ([A][ ((J [[A]])) ])
            ([A]   (J [[A]])  )                 clarify
            (                 )  ☞  1           cancel
```

Multiplication has a different identity than addition because the inverse of division is an exponential power relation. The James form `((J [[A]]))` can be read as a negative power, $\#^{-\log_\# A} = 1/A$. Were it not the case that anything to the zero power is defined to be 1, the identity for multiplication would not be 1.

For **power/logarithm**, the James inverse is function-like. Using embedded bases, (A) is a power and [A] is the corresponding logarithm. The inverse relation is defined by the Inversion axiom, and is structurally independent of the base.

$\log_\# \#^A = A$ ☞ `[(A)] = A`

$\#^{\log_\# A} = A$ ☞ `([A]) = A`

Alternatively, we can use the form of power and the form of logarithm with a *specific* base.

```
B^A       ☞   (( [A]        [[B]]    ))
log_B A   ☞    ([[A]] (J [[[B]]]) )
```

The form of the power/logarithm function inverse is fundamentally different since the nesting levels in the logarithm are not balanced. The logarithm is the only conventional function with the property of unbalanced frame structure.

```
B^(log_B A)  ☞  (([([[A]] (J [[[B]]]))] [[B]]))
                ((  [[A]] (J [[[B]]])   [[B]]))          clarify
                ((  [[A]]                    ))          cancel
                      A                                  clarify

log_B B^A  ☞  ( [[ (([A][[B]])) ]] (J [[[B]]]) )
              (       [A][[B]]      (J [[[B]]]) )        clarify
              (       [A]                       )        cancel
                       A                                 clarify
```

It is the base rather than the power that participates in Reflection. From these power/logarithm structures we can see why James computations are base-free. In each case, the base reduces to *void*. Whether the base is represented as B or as # does not matter.

For **power/root**, the inverse structure is that of multiplication, but at the level of exponents. Multiplication/division can be seen as addition of exponents, similarly power/root can be seen as multiplication of exponents. The power/logarithm inverse relationship applies Reflection to the *base* of each. In contrast, the power/root pseudo-inverse relation applies Reflection to the *power* of each. Again we can examine the James forms.

```
#^A     ☞   ((       [A]     [[#]]    )) = (       A     )
#^(1/A) ☞   (( (J [[A]]) [[#]]        )) = (((J [[A]])))
```

Here is the power/root pseudo-inverse:

```
(#^A)^(1/A)  ☞   (( [((J [[A]]))] [[(A)]]  ))
                 ((    (J [[A]])   [ A ]   ))            clarify
                 ((                        )) = #        cancel
```

```
(#^(1/A))^A  ☞  (([A] [[ (((J [[A]]))) ]] ))
                (([A]       (J [[A]])      ))      clarify
                ((                         )) = #  cancel
```

We can see that the reflection [A](J [[A]]) is identical to that of multiplication, which means that power/root does *not* identify a different type of inverse relationship. We might say that the *object inverse* of a power is a logarithm, while the *operator inverse* of a power is a root.

J-frames

J is a *new type of numeric unit*, in the same category as 1, –1, ±1 and ±i. The *unit* J is a **pure object**, not an operation. It is the J-frame that converts J into an operation. J ☞ $\log_{\#} -1$

The form of J can be deconstructed to express the real unit 1 in terms of J. Exploring these relations leads to some fascinating forms such as the void-equivalence of J J.

–1 ☞ (J [o]) = (J) ☞ $\#^{J}$

1 = – –1 ☞ (J [(J)]) = (J J) ☞ $\#^{J\ J} = \#^{J} \times \#^{J}$

(J J) = ( )

J J = *void*

J = [(J)] ☞ $\log_{\#} -1$

J J ☞ $(\log_{\#} -1) \times (\log_{\#} -1) = \log_{\#} -1^{2} = \log_{\#} 1 = 0$

The J-frame is similar to the magnitude frame but instead of magnitude it manages polarity.

$(power\ [value])_{base}$ ☞ $value \times base^{power}$

$(J\ [value])_{base}$ ☞ $value \times base^{J} = value \times -1$

The J/2 Frame

The J/2-frame is a special case of a J-frame with a fractional frame type of J/2. For reading convenience, I'm using the abbreviation J/2 for the form of J divided by 2: ([J] (J [[o o]])).

$(J/2\ [content])_{base}$ ☞ $content \times base^{J/2}$

This frame specifies the relation between the unit 1 and the unit i. The form of i is direct:

```
i =label= (-1)^1/2  ☞  (([[(J)]] (J [[o o]])))
                       (([  J  ] (J [[o o]])))
                       (        J/2          )  ☞  base^J/2
```

J leads to the realm of imaginary numbers and provides a new perspective on Euler's identity.

Dominion

This void-equivalent theorem is the ground case of the mechanism that permits replication of structure without changing cardinality. It permits the introduction of any numeric form A in the context of the non-numeric form [ ], thus providing our first perspective on the nature of [ ]. The square-bracket emits indeterminate numeric forms.

Dominion clearly mixes numeric with non-numeric forms, making it a chimera from the perspective of conventional expressions. Although of fundamental importance, Dominion is a *theorem* arising from the three axioms that themselves give no hint of transgressing into non-numeric interpretations. Non-numeric forms are a structural consequence of exponential processes. Here is a purely structural demonstration of Dominion. I'll use the cardinality frame that we have been interpreting as multiplication.

```
([A][ ])
([A][ ])    A      (J [A])        create
([A][ ]) ([A][o]) (J [A])        enfold
         ([A][o]) (J [A])        collect
            A      (J [A])        clarify
                void              cancel
```

6. Computation

Figure 5 shows the boundary form of several types of numbers in common use. The angle bracket abbreviation is used for the J-frame (J [...]) both in Figure 5 and in Figure 6. Although the abbreviations are convenient for reading, the main observation is that all of the common types of numbers can be expressed with configurations of the two types of boundary containers in conjunction with J.

number type	*standard expression* ☞	*James form*
zero	0	
natural number	3	ooo
negative number	–3	<3>
unit fraction	1/3	(<[3]>)
rational	3/4	([3]<[4]>)
other types of conventional numbers		
mixed number	2 3/4	2 ([3]<[4]>)
prime factors	$3^2 \times 5$	([5]([[3]][2]))
explicit base-2	1101	([2 o][2][2]) o
implicit base-2	1101	(((o) o)) o
irrational	√3	(([[3]]<[2]>))
e, *transcendental* base-e	e	(o)
imaginary and non-numeric forms		
J, *James imaginary*	log–1	[<o>]
i, *imaginary*	√–1	(([J]<[2]>))
complex	4+3i	4 ([3]([J]<[2]>))
pi, *transcendental*	π	([J]([J]<[2]>))
negative infinity	–∞	[ ]
positive infinity	∞	<[ ]>

Figure 5: *Types of numbers*

In Figure 6 the common operations of arithmetic are expressed as applications of the James axioms, providing a uniform computational model for half-a-dozen apparently different operations. The cost is *many small but similar transformations*. The gain is to be able to see that the diversity in computation rests only on Inversion, Arrangement and Reflection. Figure 6 uses the explicit unit form of natural numbers (rather than their digit abbreviations) in order to illustrate the fine-grain details of each computation.

expression	☞	*James form and its reduction*	
4 + 2	☞	oooo oo	
6	☜	oooooo	put
4 − 2	☞	oooo <oo>	
2	☜	oo	cancel
4 x 2	☞	([oooo][oo])	
		([oooo][o]) ([oooo][o])	disperse
8	☜	oooo oooo	clarify
4 ÷ 2	☞	([oooo]<[oo]>)	
		([oo]<[oo]>) ([oo]<[oo]>)	disperse
2	☜	( ) ( )	cancel
2 ÷ 4	☞	([oo]<[oo oo]>)	
		([oo]<[([oo][o])([oo][o])]>)	enfold
		([oo]<[([oo] [oo])]>)	collect
		([oo]< [oo] [oo] >)	clarify
		([oo]< [oo]> <[oo] >)	split
1/2	☜	(<[oo] >)	cancel
4^2	☞	(([[oooo]][oo]))	
		(([[oooo]][o]) ([[oooo]][o]))	disperse
		([oooo] [oooo])	clarify
		([oooo][o])([oooo][o])([oooo][o])([oooo][o])	disperse
16	☜	oooo oooo oooo oooo	clarify
$\sqrt{4} = 4^{1/2}$	☞	(([[oo oo]]<[oo]>))	
		(([[([oo][o])([oo][o])]]<[oo]>))	enfold
		(([[([oo] [oo])]]<[oo]>))	collect
		(([[oo] [oo]]<[oo]>))	clarify
		(([([[oo]][oo])]<[oo]>))	collect
		(([[oo]][oo] <[oo]>))	clarify
		(([[oo]]))	cancel
2	☜	oo	clarify

Figure 6: *Examples of James arithmetic calculation*

Demonstration

Here are two examples of using J for numeric reduction:

```
(A^-1)^-1  ☞   ((J [[ ((J [[A]] )) ]]))
               ((J      J [[A]]      ))            clarify
               (          [A]         )            J J =
                           A         ☞  A          clarify
```

```
        (A + 1)(A - 1) = A^2 - 1   ☞

    ([A o][A (J)])
    ([A][A]) ([o][A]) ([A][(J)]) ([o][(J)])        disperse
    ([A][A])      A    ([A]  J  )      (J)         clarify
    ([A][A])                           (J)         cancel
                                    ☞   A^2 - 1
```

7. Correspondence with Laws of Form

Logic adds a new type of unit to compare to the James units ( ) and [ ]. I'll use a new typographic delimiter, 〈 〉, rather than Spencer Brown's logic boundary ⏋ in this section in order to maintain visual comparability. The significant difference between logic and numeric boundaries is the necessity to suppress the powerful void-based logic reduction axioms.

All units maintain their existence:

```
        ( ) ≠              [ ] ≠              〈 〉 ≠
```

One James unit accumulates and the other is the same as a logic unit. Logical Truth appears to be a type of numeric infinity.

```
        ( )( ) ≠ ( )       [ ][ ] = [ ]       〈 〉〈 〉 = 〈 〉
```

Nested pairs do not condense, just like the logic unit.

```
        (( )) ≠ ( )        [[ ]]  undefined   〈〈 〉〉 ≠ 〈 〉
```

The **Inversion** axiom eliminates mutual containment of the two types of numeric unit, as it does for the logic unit.

```
        ([ ]) = [( )] =              〈〈 〉〉 =

        ([A]) = [(A)] = A            〈〈A〉〉 = A
```

The **Arrangement** axiom is the same for both numeric inversion frames and for the logic boundary.

(A [B C]) = (A [B]) (A [C]) ⟨A ⟨ B C⟩⟩ = ⟨A ⟨B⟩⟩ ⟨A ⟨C⟩⟩

Dominion too corresponds to a version of Spencer Brown's Integration theorem:

([ ] A) = *void* ⟨⟨ ⟩ A⟩ = *void*

Reflection then is the only James axiom that differs significantly from *Laws of Form,* although it does bare a structural resemblance to both the axiom of Position and the Occultation theorem

A (J [A]) = *void* ⟨A ⟨A⟩⟩ = *void* A ⟨B ⟨A⟩⟩ = A

The other substantive difference between arithmetic and logic is that the logic boundary is semipermeable. **Pervasion** asserts that replicas of outside forms can cross a logic boundary to the inside, something that is forbidden for any type of numeric boundary. Semipermeability gives limited permission to replicate without changing cardinality. This is not surprising since logic does not support accumulation of cardinality.

8. Multidimensional Forms

It is not the concepts represented by the James language that are multi-dimensional, it is the language itself that has different dialects or "notations" expressed in different dimensions. One implication is that a series of transformations can be *animated*. Another implication is that many transformations can occur concurrently, all at the same time.

A contains B

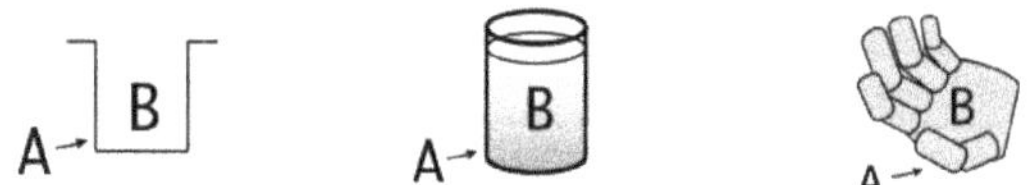

symbolic *iconic* *concrete* *experiential*

Boundary languages are *visceral*. Interpretation will remain constant as the boundary representation is transcribed across dimensions, from 1D strings to 2D icons to 3D architectures to 4D temporal experiences. There is no abstract/concrete dichotomy, so that boundary languages are much easier to understand. No mind/body split, so boundary forms are much easier to tolerate. In contrast, string encoding cannot be experienced, it must be learned via memorization. Consequently string languages remain necessarily *cerebral*.

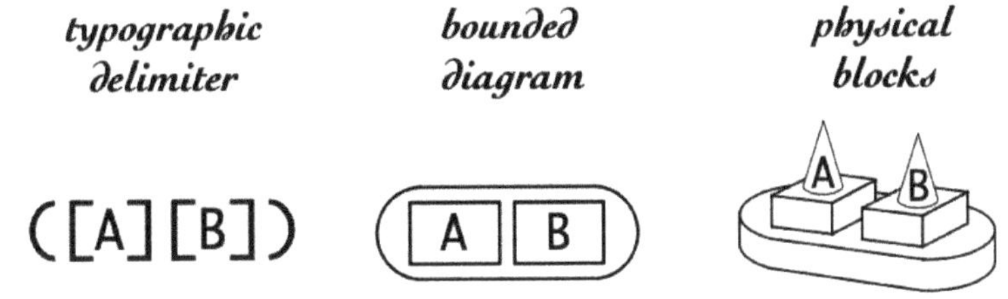

Figure 7: *One-, two- and three-dimensional forms of multiplication*

It is a distinct advantage to represent mathematical concepts across many different formats, not only symbolically but also diagrammatically, physically and experientially. The printed page limits representation to symbolic and iconic forms, but by projecting volumetric forms onto paper, we can approximate concrete and experiential languages. The image of a box can elicit imagination of a box. Leibniz: "The best signs are images; and words, insofar as they are adequate, should represent images accurately."[9]

Dialects

Containment relations themselves can be expressed not only as configurations of containers, but also as maps, networks and symbolic equations. Most string languages can also be expressed as spatial networks. A difference, though, is that a James icon embodies its operational semantics. The container boundary is the only diagrammatic component we need. It visually and computationally preserves the dependencies of containment, which itself can be interpreted as nesting, sequence, stacking, connectivity and several other types of *physical* relationship between container and contained, as illustrated in Figures 7 and 8.

Figure 7 shows the James form of multiplication expressed as one-, two-, and three-dimensional containers.

- The **string** dialect is digital and encodable. The language consists of one-dimensional delimiters in fractured bracketing relationships with one another.
- The **bounded** dialect shows two-dimensional containment. The language consists of nesting relationships among closed curves.

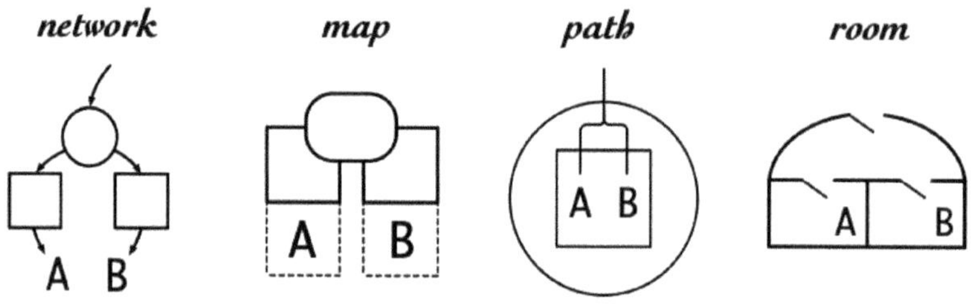

Figure 8: *More forms of multiplication*

- The **block** dialect is manipulable. The three-dimensional language consists of physical objects in stacking relationships with one another.

Figure 8 shows James multiplication in some other spatial dialects.

- The **network** dialect is a traversable acyclic graph. The language consists of nodes and links displayed in two- or three-dimensions.
- The **map** dialect is a traversable territory. The language consists of areas with shared borders and an explicit *you-are-here* starting point.
- The **path** dialect shows border crossings over a boundary. The language consists of a directed path crossing singular boundaries.
- The **room** dialect is a three-dimensional environment inhabited by a participant. The language is arrangements of rooms and doors.

9. Summary

Perhaps the most challenging new idea is that many of the conventional *concepts* within arithmetic and algebra are unnecessary in order to conduct the operations of arithmetic and algebra. Commutativity and associativity are interpreted as sequential concepts that are not relevant to a spatial, parallel form of arithmetic.[10] There is no ordering imposed on the contents of a container. Grouping is defined solely by containment. Unlike functions which specify a precise number of arguments, a container can accommodate any number of contents. There is no concept of *arity*. The `0` null object of addition and the `1` null object of multiplication are opposite sides of the same distinction, 1 from the outside and 0 from the inside. James algebra eliminates three of the five fundamental properties of algebraic groups. From the James perspective only inverse and distribution are fundamental.

James algebra may not be initially friendly to the eye, primarily due to its unfamiliarity. It becomes friendly to the mind only after exerting some effort to achieve familiarity. Familiarity can be achieved cognitively or iconically or physically. We cannot expect to easily recognize a radical revision of the arithmetic of numbers that has been standardized across the globe for a century. We can also not easily shed the perspectives learned in twelve years of grade school and high school numeric math courses. Or, if you choose, simply adopt the mind of a novice, someone who has never encountered a number greater than two.

Mathematical concepts themselves are defined by containment patterns. We do not have to abandon the familiar concepts of symbolic math, since carrying them into a reading of a James form at least provides comfort in familiarity. The adventure though is in finding other concepts, *iconic concepts*, overlaid and interacting with our familiar concept of number. It is these new perspectives that provide the motivation to explore. Similar to Conway numbers[11] (aka surreal numbers), this exploration is not yet pragmatic; the goal is simply to shed some new light on what it means to be a number. A deeper objective is to show that symbolic arithmetic as we know it is not fundamental, but rather just one of many formal approaches to magnitude.

Boundary math does not seek novelty as a goal, it seeks simplicity. Most usually, the simplest path is both invisible and obvious. Simplicity is gained not by a higher level of abstraction but by embracing *only one thing*, the distinction, and only one operation to actualize distinction as computation, containment.

The [ ] unit in combination with J creates a cacophony of non-numeric structures including the exotic and indeterminate expressions of conventional math such as negative infinity, infinitesimals, divide-by-zero and square-root-of-negative-one. This article has tried to avoid those creatures.

J is a *newly resurrected additive imaginary unit*, from which the multiplicative i, √-1, can be constructed. It can be interpreted as $\log_{\#} -1$. J leads to new perspectives on trigonometric functions, calculus derivatives and imaginary spaces.

Endnotes

1. **opening quote:** B. Mazur (2003) *Imagining Numbers.* Picador. p.163. Mazur continues on p.166: "It is also easy to underestimate the difficulties of comprehension that any change of notation presents."

2. **a three volume series *Iconic Arithmetic*:** W. Bricken *Iconic Arithmetic* (2019) Vol. I: The Design of Mathematics for Human Understanding; (2019) Vol. II: Symbolic and Postsymbolic Formal Foundations; (2021) Vol. III: The Structure of Imaginary and Infinite Forms. Unary Press. Available online at https://www.amazon.com/Iconic-Arithmetic-Design-Mathematics-Understanding/dp/1732485135, https://www.amazon.com/Iconic-Arithmetic-II-Postsymbolic-Foundations/dp/1732485143, https://www.amazon.com/Iconic-Arithmetic-III-Structure-Imaginary/dp/1732485151.

3. **likewise be out of the contained:** L. Euler (1802) Letter CIV Different Forms of Syllogisms, (2/21/1761), H. Hunter (trans.) *Letters of Euler to a German Princess.* Gale Ecco. p.406.

4. **arranged in sentences and equations:** H. Simon, forward to B. Chandrasekaran, J. Glasgow, and N. Narayanan, (eds.) (1995) *Diagrammatic Reasoning.* MIT Press. p.xi.

5. **while the sentential representation does not:** J. Larkin & H. Simon (1987) Why a diagram is (sometimes) worth ten thousand words. In Chandrasekaran *et al*, p.696.

6. **even the most trivial consequence needs to be inferred explicitly:** J. Barwise & J. Etchemendy (1996) Visual information and valid reasoning. In G. Allwein & J. Barwise (eds.) *Logical Reasoning with Diagrams.* Oxford University Press. p.23.

7. **any base is as good as any other:** P. Lockhart (2012) *Measurement.* Harvard University Press. p.387.

8. **this conventional form is as simple as it gets:** Log does distribute across multiplication, a fundamental feature that allows logarithms to convert multiplication into addition. Addition itself is the distribution of logarithms across multiplied exponents.

9. **words, insofar as they are adequate, should represent images accurately:** Leibniz to Tschirnhaus, end of 1679 (Math., IV, 481; Brief., I, 405), as quoted in L. Couturant *The Logic of Leibniz,* Ch. 4 footnote 93. Online 4/24 http://philosophyfaculty.ucsd.edu/faculty/rutherford/Leibniz/couturatcontents.php

10. **that are not relevant to a spatial, parallel form of arithmetic:** Peirce appears to be the first logician/mathematician to suggest that commutativity and associativity are secondary rather than axiomatic concepts.

> (4.374) Operations of commutation, like xy therefore yx, may be dispensed with by not recognizing any order of arrangement as significant. Associative transformations, like (xy)z therefore x(yz), which is a species of commutation, will be dispensed with in the same way; that is, by recognizing an equiparent as what it is, a symbol of an unordered set.

C.S. Peirce (1931-58) *The Collected Papers of Charles Sanders Peirce*. Electronic edition reproducing Vols. I-VI C. Hartshorne and P. Weiss (ed.) (Harvard University Press, 1931-1935), Vols. VII-VIII A. W. Burks(ed.) (1958). Online 3/24 https://edisciplinas.usp.br/pluginfile.php/5165117/mod_resource/content/0/The%20Collected%20Papers%20of%20Charles%20Sanders%20Peirce%20%282904s%29.pdf

11. **Similar to Conway numbers:** J. Conway (1976) *On Numbers and Games*. A.K. Peters.

PRIMES BETWEEN SQUARES: COMMENTARY ON APPENDIX 8 OF *LAWS OF FORM*

J. M. FLAGG

LOUIS H. KAUFFMAN
Department of Mathematics, Statistics, and Computer Science,
University of Illinois, Chicago
and
International Institute for Sustainability with Knotted Chiral Meta Matter,
WPI-SKCM2, Hiroshima University,
1-3-1 Kagamiyama, Higashi-Hiroshima, Hiroshima 739-8526, Japan
loukau@gmail.com

DIVYAMAAN SAHOO
Department of Electrical and Computer Engineering,
University of Massachusetts Dartmouth,
and
Graduate Program in Acoustics,
The Pennsylvania State University
divyamaansahoo@gmail.com

The authors are indebted to Andrew Crompton, Graham Ellsbury, Thomas Wolf, and Moshe Klein for insights and conversations.

An argument follows.

James Joyce, *Finnegans Wake*

Hail Thoth, architect of truth, give me words of power that I may form the characters of my own evolution. I stand before the masters who witnessed the genesis, who were the authors of their own forms, who rolled into being, who walked the dark circuitous passages of their own becoming, ...

Hail Thoth, architect of truth, give me words of power that I may intuit the symbols of dream and command my own becoming. I stand before the masters who witnessed the working of magic, ...

Hail Thoth, architect of truth, give me words of power that I may complete my story and begin life anew. I stand before the masters who witnessed the plowing of earth, who saw the seed that entered the fields spring into corn and barley, who sent the flood and sun, who saw men among the wheat swinging scythes, who saw women baking bread. ...

Hail Thoth, architect of truth, give me words of power that I may tell the truth of my own becoming. I stand before the masters who witness the judgment of souls, who sniff out the misdeeds, the imperfections, the lies and half-truths we tell ourselves in the dark. ...

Normandi Ellis, *Awakening Osiris*

Abstract

There are two conjectures regarding the distribution of primes between squares attributed to Legendre, a weak and a strong conjecture. The weak conjecture[1] is that there is *at least one* prime between consecutive squares. The strong conjecture is that there exist *at least two* primes between consecutive squares. In Appendix 8 of *Laws Of Form* (Bohmeier Verlag, revised sixth English edition, 2015), Spencer-Brown sets out to prove the stronger of the two. In unpublished material, he provides stronger and sharper theorems and conjectures on the regular and predictable increase of the number of primes between consecutive squares. In Appendix 7 of *Laws Of Form* (Bohmeier Verlag, 2015), and in *Gesetze der Form* (Bohmeier Verlag, 1997), he claims:

> The total number $t(n)$ of primes between n^2 and $(n+1)^2$, where n is a natural number greater than 1, is within the limits
>
> $$A - (B - 1) < t(n) < A + (B - 1),$$
>
> where A is $\frac{n}{\log n}$ and B is $\frac{A}{\log A}$.

In other words, the number of primes between squares is asymptotic to $\frac{n}{\log n}$. This article provides commentary and context to the number theory of Spencer-Brown.

[1]When reformulated, this conjecture is the fourth problem of Edmund Landau, who in his lecture "Gelöste und ungelöste Probleme aus der Theorie der Primzahlverteilung und der Riemannschen Zetafunktion" at the International Congress of Mathematicians in Cambridge, England, on August 23rd, 1912, characterised his four problems as "unattackable at the present state of science". In Landau's words, "Liegt zwischen n^2 und $(n+1)^2$ für alle positiven ganzen n mindestens eine Primzahl?" (Is there at least one prime number in between n^2 and $(n+1)^2$ for all positive whole n?)

0 Introduction

This paper provides a commentary and guide to Appendix 8 of *Laws Of Form*, which is an article by Spencer-Brown on number theory and his proofs of the conjecture that there are at least two prime numbers between any consecutive squared numbers n^2 and $(n+1)^2$. For example, $1^2 = 1$ and $2^2 = 4$ and the numbers between them are 2 and 3 thus we have two prime numbers between 1^2 and 2^2. This phenomenon continues for all pairs n^2 and $(n + 1)^2$ and was a conjecture in number theory since Legendre. In Spencer-Brown's appendix he gives his proofs of the conjecture. We shall see that those proofs are a highly original mixture of standard rigorous arguments and also some stated facts about the way numbers behave that would be considered conjectures by most number theorists. These phenomena are very interesting and constitute a deep observation about the nature of number itself. We hope that our guide will enable the reader to gain insight into Spencer-Brown's point of view.

Section 1 defines the square segment and introduces the stroke notation of counting primes in segments. Here, we introduce the sharp conjectures made by Spencer-Brown on the distribution of primes between squares, beginning with Theorem 1, which is a reformulation of Legendre's weak conjecture that there exists one prime between squares. The fundamental observation is that the primes $< n$ determine the primes $< n^2$, and this guides Lemmas 1 and 2, which are both proved rigorously.

Section 2 lays the foundation of Spencer-Brown's approach to the Sieve of Eratosthenes by clarifying his use of the word "strike" as an operator and operation, culminating in the "prime paradigm", which is a repeating pattern of composites generated by a set of primes. This pattern has a four-fold symmetry about multiples of the primorial (product of primes) and multiples of half the primorial.

Section 3 defines a sect as a sequence of consecutive composites and reformulates Legendre's conjecture (Theorem 1) by asking the question: what is the maximum sect that can appear in a segment? Two conjectural lemmas that are closely related to Theorem 1 are introduced here, namely Lemma 3, which asserts that maximal odd sects assume one of two forms, quadrantic or reflexive, and Lemma 4, which asserts that the quadrantic sect length is always lesser than or equal to the reflexive sect length. This section ends with a rigorous proof of Theorem 1, however, this proof rests on the validity of Lemmas 3 and 4.

Section 4 generalizes Theorem 1 to a powerful family of conjectures of the following form: the next prime after p is at most $\frac{1}{k}\sqrt{p}$ away, where $k = \frac{1}{2}$ and for primes $p = 7, 113, 1327, ...$, the value of k jumps to a larger value from which it cannot slip back. This is called the "ratchet" phenomenon. The section ends with two of Spencer-Brown's stronger conjectures: a Corollary, which states that for any $n \geq 131$ there is at least one prime in $(n - \sqrt{n}, n)$ and $(n, n + \sqrt{n})$, and Theorem 2, which is a reformulation of Legendre's strong conjecture that there exist at least two primes between squares.

Section 5 introduces Spencer-Brown's theory of elementals and connects it with his theory of modulators, rejoining his algorithm in Lemma 3 to the initials of *Laws Of Form*.

Finally, the paper ends with three appendices:

Appendix 1 provides unpublished excerpts of Spencer-Brown on sharp bounds and estimates to $\pi(x)$.

Appendix 2 contextualizes Spencer-Brown's conjectures in a set of related conjectures and theorems on the distribution of prime numbers.

Appendix 3 illustrates the sole exception to Spencer-Brown's Lemma 3 at 23.

Following Petzold's convention in *The Annotated Turing* (2008), this article uses a box when quoting excerpts from Spencer-Brown in Appendix 8 of *Laws Of Form* (Bohmeier Verlag, 2015) accompanied by elucidatory annotations and corrections.

1 Preliminaries

Appendix 8 “Primes between squares”, *Laws Of Form*, begins with the definition of “square segment” or “segment” as an interval between squares whose roots differ by 1. Hence, seg(x) indicates the stretch of natural numbers between x^2 and $(x+1)^2$. Call the square segment seg(n) when the roots are natural numbers. For example:

$$\text{seg}(2) = \{5, 6, 7, 8\},$$
$$\text{seg}(\sqrt{2}) = \{3, 4, 5\}.$$

Note that seg(2) is the set of natural numbers between 2^2 and 3^2, i.e., between 4 and 9, and seg($\sqrt{2}$) is the set of natural numbers between $(\sqrt{2})^2$ and $(\sqrt{2}+1)^2$, i.e., between 2 and ≈ 5.8284. Note, seg(0) = $\{\}$, the stretch of natural numbers between 0^2 and 1^2.

Adapting Landau’s notation $\pi(x)$ for the number of primes $\leq x$, Spencer-Brown defines $^\circ\pi\text{seg}(x)$ read “stroke pi seg of x”[2] or, equally often, “pi seg of x” to denote the number of primes in seg(x). So, $\pi(x)$ goes to $^\circ\pi(x)$. At the outset, Spencer-Brown claims:

---

[2] Note: the $^\circ$ or ‘stroke’ notation is used before by Spencer-Brown in the context of modulators (reductors), Chapter 11, *Laws Of Form*, page 66—“Let a marker be represented by a vertical stroke, thus |”—and in Spencer-Brown’s unpublished manuscript, “Introduction to Reductors”, Section 1.6: Conventions, page 10—“The standard *drive* D to a four-stroke reductor is taken to be 1010, and each unit is called a *stroke*, written $^\circ 1, ^\circ 2, ^\circ 3$, or $^\circ 4$”. So far, Spencer-Brown’s earliest known use of the term “stroke” is in “The Universal Operators in Logic”. In this unpublished manuscript, he examinines other primitive mathematical operators, of Sheffer, Stamm, Peirce, Nicod, et alia, all of which he deems inadequate due to restrictions and constraints delimiting scope of operations. He raises the question why the bar (a bar drawn over an algebraic equation) had not been considered as a “sole” operator and writes that the theoretical significance of its practical form and applicability were missed. The bar just needed some terminal modification so that it could be read correctly. He says, “To overcome the mechanical difficulty, it is suggested that the bar should be determined by a stroke, thus, ⌉”. It is also interesting to note that the Theory of Types, whether simple or ramified, uses right upper superscripts to denote a proposition’s type and place in an elevated hierarchy to elide and avoid paradox, thus eliminating any formal system incapable of referring to itself. Spencer-Brown possibly adapted this notation by moving the superscript to the left of the operator/operand and designating it as a circle (or ouroboros) of modular degree with valid closure, referring to and calculating its contents recursively, and calling it a “stroke.”

i. When x is a real number > the least absolute value of $(\sqrt{2} - 1)$, $°\pi\text{seg}(x) \geq 1$.
ii. When x is a natural number n, $°\pi\text{seg}(n) \geq 2$.
iii. When $n \geq 6$, $°\pi\text{seg}(n) \geq 3$, and the minimum number of such primes, or more exactly its lesser bound, will increase regularly and predictably as n increases.

In this context, Spencer-Brown is referring to his Prime Limit Theorem, Equation (5) on pg. 182, Appendix 7, *Laws Of Form* (Bohmeier Verlag, revised sixth English edition, 2015), that $°\pi\text{seg}(n)$ is asymptotic to $\frac{n}{\log n}$.

After defining the (square) segment, Spencer-Brown introduces Theorem 1, which is a reformulation of Legendre's weak conjecture.

> **Theorem 1.**
> There is at least one odd prime in the square segment of every natural number.

Spencer-Brown proves Theorem 1 by introducing the following Lemmas 1 through 4. *Note: Lemma 1A is not provided by Spencer-Brown but is included here for clarity.*

> **Lemma 1**:
> A number h is prime iff h is an integer and h does not make an integer quotient with any integer i in case $2 \leq i \leq \sqrt{h}$. [a]
> **Proof:** There is no such i unless $\sqrt{h} \geq 2$. If i divides h then $h/i = j$, say. Now either $i = j$, $i < j$, or $i > j$. In the first case $i = \sqrt{h}$, in the second $i < \sqrt{h}$, and in the third $j < \sqrt{h}$ and can replace i as the divisor. So effectively all proper divisors of h will have been tried by confining i in case $2 \leq i \leq \sqrt{h}$. □
>
> [a] The constraint on i ensures that i, $\sqrt{h}$, and h are all positive, but in fact primes, like moduli, are signless.

In the footnote to Lemma 1, Spencer-Brown is referring to the modular number system. This powerful Lemma states that in order to test whether any number is a prime, it is sufficient to look for factors only up to the square-root of that number. This fact is well known.

NOTE: The following lemma is not provided by Spencer-Brown.
Lemma 1A:
The least prime factor of any odd composite between n^2 and $(n+1)^2$ is $\leq n$.

Proof: The primes $p_1, p_2, p_3, \ldots, p_j$ solely determine all the primes up to p_{j+1}^2 (see Volume I, Chapter XIII, *History of the Theory of Numbers*, L. E. Dickson). If $n = p_j$, then $(n+1)^2 < p_{j+1}^2$. If $(n+1) = p_{j+1}$, then $(n+1)^2 \leq p_{j+1}^2$. In both cases, the primes $p_1, p_2, p_3, \ldots, p_j \leq n$ determine all the primes up to $(n+1)^2$, so the least prime factor of any odd composite between n^2 and $(n+1)^2$ is $\leq n$. □

Lemma 2:
There are n odd numbers between n^2 and $(n+1)^2$.

Proof: $(n+1)^2 - n^2 = 2n+1$. In a stretch of $2n+1$ integers, n must be odd. □

First consider, seg(1) = $\{2,3\}$, seg(2) = $\{5,6,7,8\}$, seg(3) = $\{10,11,12,13,14,15\}$, and seg(4) = $\{17,18,19,20,21,22,23,24\}$. Observe the odd numbers in seg(5):

5^2	27	29	31	33	35	6^2
□	•	•	•	•	•	□
5	3			3	5	

The vertical line marks the primorial $2 \times 3 \times 5$. The numbers 5 and 3 in the third line indicate the first prime that divides the odd number above it, so 29 and 31 are prime. By Lemma 2, there are 5 odd numbers between 5^2 and 6^2.
By Lemma 1A, the least prime factor of any odd composite in seg(5) is ≤ 5.
By Lemma 1, $h = 29$ is prime since it does not make an integer quotient i with any integer i in case $2 \leq i \leq \sqrt{29} \approx 5.385$, i.e., $i \in \{2,3,4,5\}$. Similarly, $h = 31$ is prime.

Call all primes $\leq n$ the *prime generators of* n^2 or *pg of* n^2 for short. Alternatively, the primes $\leq \sqrt{n}$ are the *prime generators* of n. This idea anticipates the recursive relationship of the prime counting function,[3] i.e., $\pi(n)$ is a function of $\pi(\sqrt{n})$. This relationship is illustrated by Spencer-Brown in Appendix 9 of *Laws Of Form* (Bohmeier Verlag, 2015).

[3]In a footnote on page 190, Spencer-Brown writes, "The true pg of n is *all* the multiplicative integers $\leq \sqrt{n}$, because it demands no previous knowledge of primes". The Legendre/Spencer-Brown formula in Appendix 9 of *Laws Of Form* shows how the prime counting function is self-referential. The formula is $\pi(n) = \pi(\sqrt{n}) + \sum_{d=1}^{\sqrt{n}} \mu(d) \left[\frac{n}{d}\right]$, where $d \in$ {the set of all combinations of prime generators $\leq \sqrt{n}$, including 1} and [.] is the floor function. For $d = 1$, Spencer-Brown changes $\left[\frac{n}{d}\right]$ to $\left[\frac{n-1}{d}\right]$. See "Laws Of Form and the Riemann Hypothesis", *Laws Of Form: A Fiftieth Anniversary*, World Scientific, 2021.

2 The Foundation

On page 190 of Appendix 8, Spencer-Brown lays the foundation to his approach:

> Imagine the system of integers, call it an arithmetic A, mapped onto a line of regularly-spaced indistinguishable marks or points stretching endlessly in either direction. Call these marks the elements of a system S.

> Consider prime multisectors $d_1, d_2, d_3, \ldots$ that can be imbedded in every second, every third, every fifth, and so on of the elements in S.

Multisector d_i is equal to p_i (the ith prime), i.e. $d_1 = 2, d_2 = 3, d_3 = 5, \ldots.$ Hence, fix an origin and mark the prime p_i (and all integer multiples of p_i) with d_i as indicated:

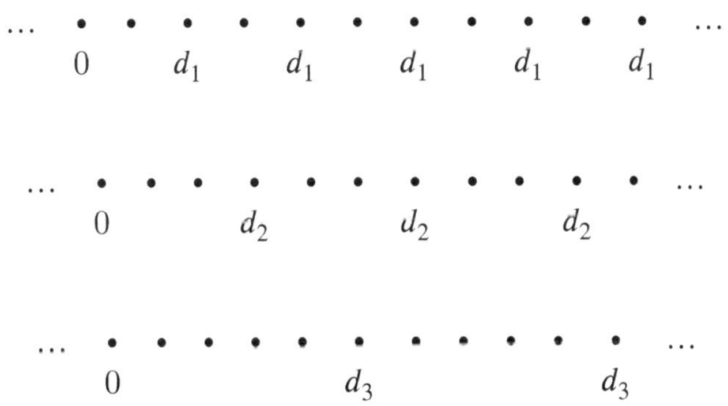

> ... the multisectors are not primes or prime divisors, but merely operators on elements of S that can be placed where we please to imbed with or strike certain of these elements.

In the handwritten edition of *Primes Between Squares*, Spencer-Brown uses the terms "multisector" and "decimator" interchangeably. He writes, "Since decimation is a multisection by ten, we may conveniently extend the meaning of this more-familiar term and call each multisector a decimator, giving it the name of the prime divisor whose span it is associated with". The following two pages excerpt page 190 of Appendix 8.

> Consider S after d_1 has struck every second element, and eliminate the elements that have been struck, leaving the remainder to form a system S', say. S' is now identical to S, and d_1 is the only multisector that makes such an identity. So we can save labour by mapping the odd numbers in A, call them A', onto the elements of S'.

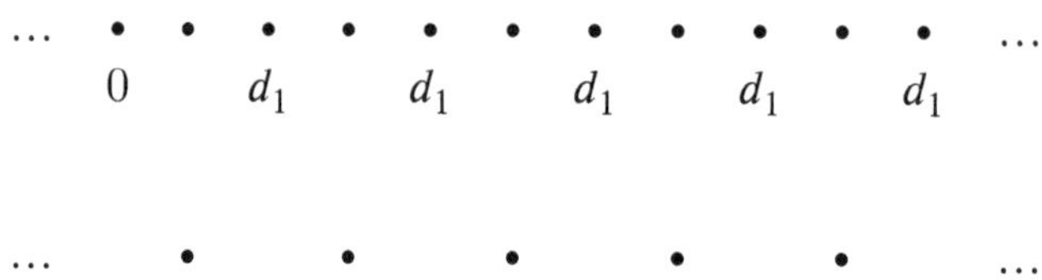

In S, the imbedding of multisector $d_2 = 3$ appears as:

And in S' (odd numbers), the imbedding of multisector $d_2 = 3$ appears as:

In fact, 0 is not a member of S' (it is not an odd number) so it is represented by $\odot$ and labeled (0) to aid the reader. This point must be erased as it does not belong in S'. When a multisector d_j is imbedded in S, the point labeled 0 must be relabeled d_j. If these rules are to be followed, then the patterns created by multisector d_2 in S and S' appear to be "identical", i.e., they are both lines of regularly-spaced indistinguishable marks or points stretching endlessly in either direction that look like ... • • d_2 • • d_2 • • d_2 • • ... . Similarly, the imbedding of multisector $d_3 = 5$ in S and S' identically look like ... • • • • d_3 • • • • d_3 • • • • d_3 • • • • ... . Verify this! In general, the imbedding of any multisector in S is identical to its imbedding in S'. Hence, S and S' are indistinguishable from one another and d_1 is the only multisector that makes this identity. In S, the sequence that starts with 0 can be regarded as starting from some other number such as d_j, so 0 as origin repeats at the primorial $p_1 \times p_2 \times ... \times p_j$, where $p_j = d_j$ is a prime, and these are what Spencer-Brown calls *paradigmatic places* or pp. What distinguishes system S from arithmetic A is that multisector imbeddings in S can be frozen and these patterns can be shifted like a ruler along the number line and put in a *nonparadigmatic place* or np. Spencer-Brown puts these in square segments.

> Every possible arrangement of multisectors in S is matched by a similar arrangement of divisors somewhere in A, and the reason for separating the concepts of S and A is to be able to experiment with the multisectors in S, by shifting them from place to place, which we cannot do with the divisors in A. In other words, I have made in S a working model of A, which we can take to pieces and play with the pieces, with the aim of discovering sharp limits to what is possible in A by seeing what is possible, and what is not, in S.

Focusing on the "imbedding" operation of multisectors, Spencer-Brown uses the word "strike" with different distinctions of use intended, as a noun, collective noun, and operator, collapsed into one by the second canon, *contraction of reference*.[4] He writes:

> We can employ the noun *strike* to indicate the appearance of a section of S after the operation of a set of d's. It will appear similar to a section of numbers in A in which a corresponding set of prime divisors have imbedded to make composites, leaving the untouched numbers therein as new primes. (Compare the sieve of Eratosthenes.)

"Strike" as a *verb*, or operator, first, is used in reference to the Sieve of Eratosthenes, an ancient algorithm used to generate prime numbers. The following is an illustration of the sieve.

[4]"Let injunctions be contracted to any degree in which they can still be followed" (Chapter 3, *Laws of Form*, Bohmeier Verlag, 2015)

Begin with a list of 20 numbers. Omit the number 1. Begin with the number 2. Strike all mutiples of 2 (except itself) off the list. Call the remaining numbers "holes". After 2, the first hole is 3. Use 3 to strike multiple of 3 off the list. After 3, the first hole is 5. Use 5 to strike multiple of 5 off the list, ... The holes that remain are all the primes up to 20. These determine all primes (holes) up to 20^2, ...

2	3	4	5	6	7	8	9	10	11	12	13	14	15	16	17	18	19	20
⓶	⓷	~~4~~	⓹	~~6~~	⓻	~~8~~	~~9~~	~~10~~	⑪	~~12~~	⑬	~~14~~	~~15~~	~~16~~	⑰	~~18~~	⑲	~~20~~

In an unpublished manuscript "Commentary on Appendix 9", Spencer-Brown produces a beautiful method of sieving prime numbers, where × marks the spot.

The sieve of Spencer-Brown

Write down the natural numbers in order beginning at 1 and ending at shall we say 20. Put zero (which is not a natural number) in front of the sequence and, counting from zero, mark every second, every third, every fourth, and so on number. The numbers with just one mark are the primes.

Illustration

0	1	2	3	4	5	6	7	8	9	10	11	12	13	14	15	16	17	18	19	20
		x	x	x	x	x	x	x	x	x	x	x	x	x	x	x	x	x	x	x
				x		x		x	x	x		x		x	x	x		x		x
						x		x		x		x		x	x	x		x		x
												x				x		x		x
												x						x		x

"Strike" as a *noun* refers to a representative piece of S or S' when a multisector (or a collection of multisectors) is (are) imbedded in S or S'. This representative piece of S in its full form is called the "full reflexive strike". In the full reflexive strike, we set a 0 point and mark both sides of 0. This yields a four-fold pattern consisting of four quadrants with two "mirrors" or lines of symmetry located at the primorial ($=p_1 \times p_2 \times .. \times p_j$) and half the primorial (which defines the 1st quadrant, or quantum, or half-strike). The following two excerpts are from page 190 of Appendix 8:

> We examine what happens when a set of relatively prime multisectors strike over a consecutive set of points in S or S'.

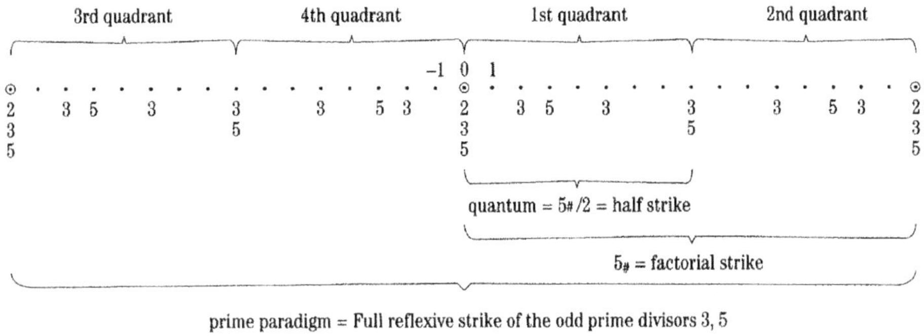

Figure 1. **Full reflexive strike of the odd primes 3, 5. The ringed points are mirror-points at multiples of 2, which do not strictly belong to the odd-strike.**

Observe in Figure 1 that we can place inwards-facing mirrors at each end of any quadrant to get the full picture to infinity, so: the quadrant is the *quantum*, or least stretch of space that contains all the visual information we require.[a] Figure 1 shows the repetitive arrangements, and the names I have given them, of a strike of two odd prime decimators or divisors. It can of course be extended to any number of such decimators or divisors.

[a]The mirror placings are of particular interest since they must be through the exact location of a point. Thus if these points are euclidean, having position but no size, they will be entirely obliterated by the mirror. To reconstruct them we must allow them an infinitesimal size, so that half the point will appear in the quadrant, and the other half as its image in the mirror.

"Strike" as a *collective noun* also refers to the prime factorial, primorial, hash, or prime generator, pg. A definition is provided by Spencer-Brown on page 191 of Appendix 8:

Write *x#* (*or* x_j), say *x*-hash (*or x-strike*), for the prime factorial of a nonnegative real number *x*. *x#* is defined as the factorial of the integer part of *x*, divided by the composite numbers $\leq x$. For example, $\sqrt{80}$# = 8.944...# = (8!/8.6.4) = 7# = 210. Thus the primes ≤ 7 form the pg of 80, their strike covers 210 integers, and their quantum 105.

3 Sects

Spencer-Brown introduces new terminology for the composite numbers that are struck. He calls a sequence of consecutive composite numbers a "sect". On page 192, he writes:

> We shall need one further special term. Mnemonic: imagine a *set* as a Show of Elements Together. Then construct the terms *sect* to denote a Show of Elements Consecutively Together. By a happy chance it extends the ordinary meaning of the word 'sect' and perfectly fits its etymology, which is from Middle English *secte*, from Old French, from Latin *secta*, from *sectus*, archaic past participle of *sequi* to follow.

3.1 Odd sects

From here on out, sects will refer to *odd* sects, sequences of consecutive *odd* composites. We return to our previous example of seg(5). Recall that a square segment, denoted seg(n) is the stretch of dots between squares, and 5 and 3 in the third line indicates our convention of writing the first prime that divides the odd number above it.

5^2	27	29	31	33	35	6^2
□	•	•	•	•	•	□
5	3			3	5	

The sect $\underset{3}{\bullet}\ \underset{5}{\bullet}$ is a sequence of consecutive odd composites in seg(5).

Another sect we see above is $\underset{5}{\bullet}\ \underset{3}{\bullet}$, although this sect is not entirely in seg(5).

The sects $\underset{3}{\bullet}\ \underset{5}{\bullet}$ and $\underset{5}{\bullet}\ \underset{3}{\bullet}$ are maximum odd sects using the pg of $5^2 = \{(2), 3, 5\}$.

The following are illustrations of the distribution of odd sects in square segments.

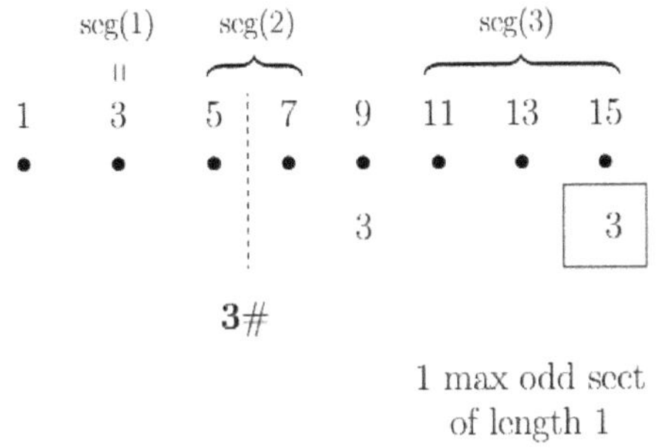
seg(1)
seg(2)
seg(3)
1 3 5 7 9 11 13 15
3
3
3#
1 max odd sect
of length 1

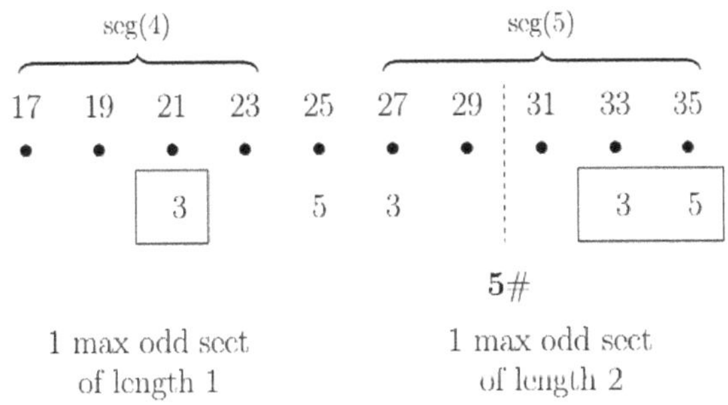
seg(4)
seg(5)
17 19 21 23 25 27 29 31 33 35
3
5 3
3 5
5#
1 max odd sect
of length 1
1 max odd sect
of length 2

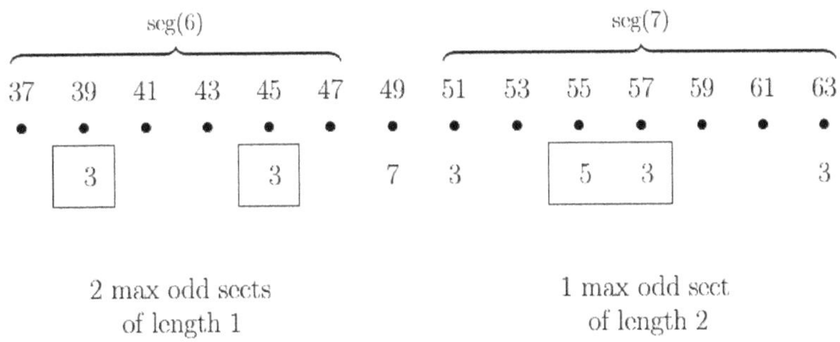
seg(6)
seg(7)
37 39 41 43 45 47 49 51 53 55 57 59 61 63
3
3
7 3
5 3
3
2 max odd sects
of length 1
1 max odd sect
of length 2

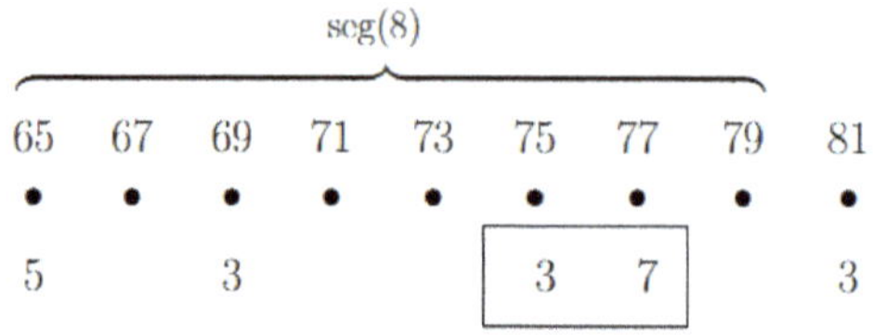

1 max odd sect
of length 2

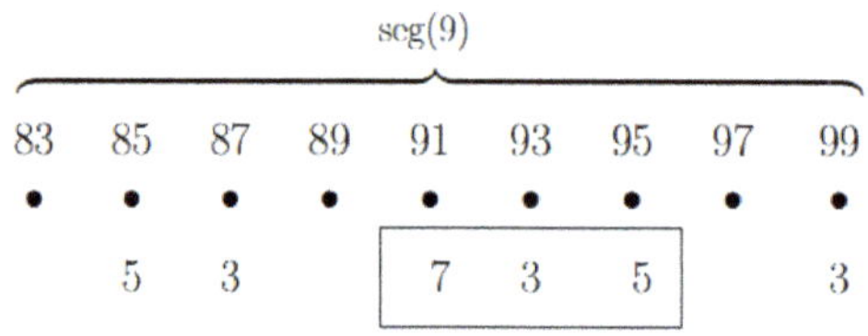

1 max odd sect
of length 3

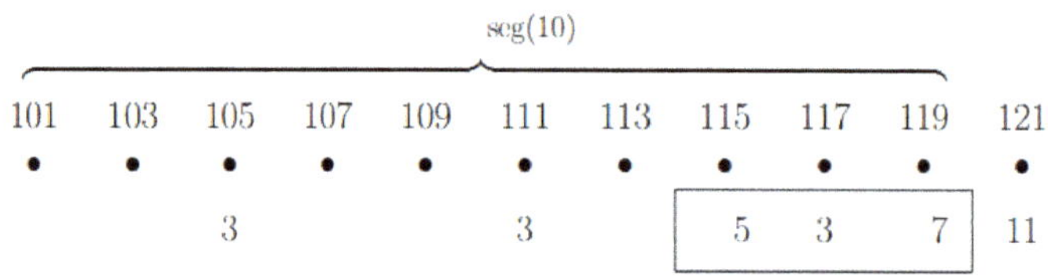

1 max odd sect
of length 3

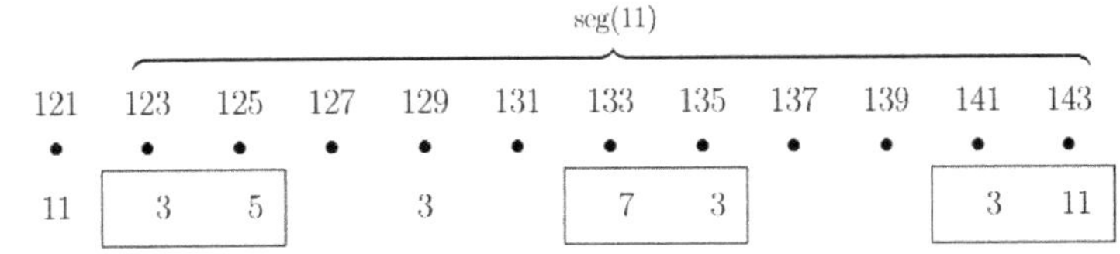

3 max odd sects
of length 2

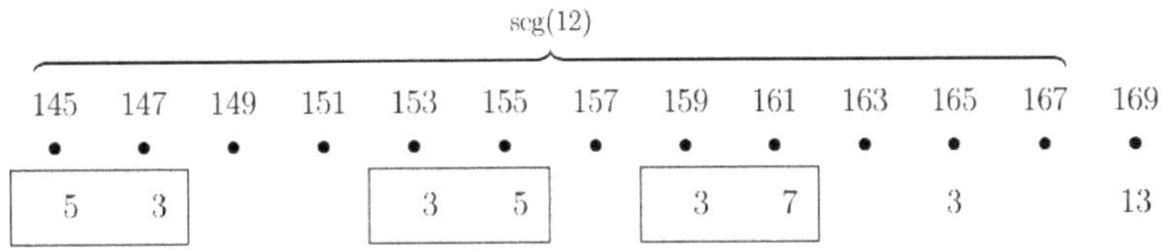

3 max odd sects
of length 2

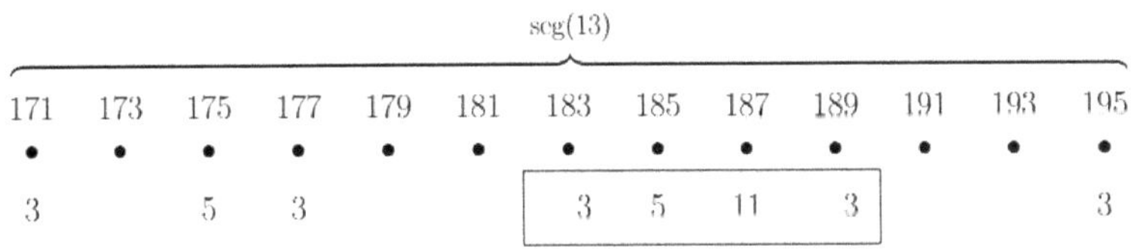

1 max odd sect
of length 4

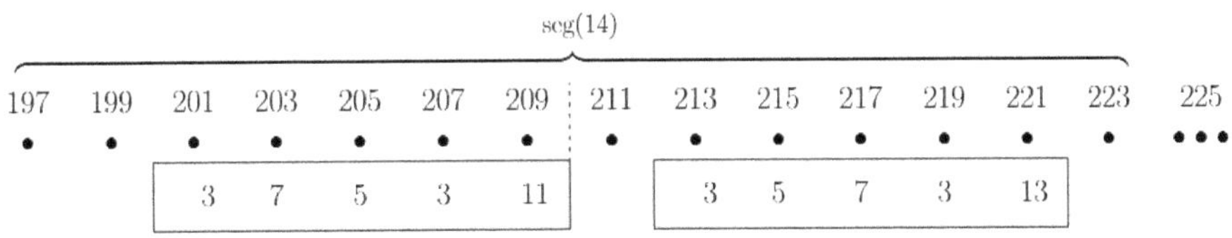

7#

2 max odd sects
of length 5

3.2 What is the maximum sect that can appear in a square segment?

Spencer-Brown proposes a **central** question that relates the lengths of maximum sects in square segments to Legendre's conjecture:

> What is the maximum set of successive composites that can be generated by the prime divisors in the pg of n^2? If the answer comes to less than $2n$, or less than n if we consider the odd numbers only, then we are done, since it will mean that not all the numbers in seg(n) can be struck by these prime divisors, so at least one of the numbers in seg(n) must be prime.

The answer to Spencer-Brown's question lies in the *prime paradigm*, i.e., Figure 1:

> From Figure 1 we see there are just two basic ways in which a maximum sect of prime divisors $\leq p$ might be constructed.
> 1. By marrying the two reflexive halves of the paradigm of prime divisors as they appear on either side of zero, and thus on either side of any multiple of $p\#$, and occupying the two holes corresponding to 1 and -1 with proper divisors, or (if possible) by some improvement on this recipe, and
> 2. By noting the natural sect that begins at 2 (or using the odd numbers, from 3 onwards) in the first quadrant, and ends just before the least prime $q > p$.
> These two recipes are together exhaustive: indeed if we are allowed to search the entire quantum of the first quadrant, instead of just the beginning of it, to which we are confined by recipe 2, we shall be certain to find the maximum sect we are looking for, or its mirror image.

Call the sect determined by recipe 1 a *reflexive* or R-sect, and that determined by recipe 2 a *quadrantic* or Q-sect. As we shall see, these recipes can be applied at the origin as well as beyond the origin, namely at multiples of the primorial, $p\#$.

Spencer-Brown provides the following notation, conflating primes p and prime multisectors d, using p and d interchangeably:

Write $p_{\sim j}$ (and similarly $d_{\sim j}$) to indicate the set of all primes (and similarly all prime multisectors) $\leq p_j$ (or d_j). In S' and the odd number arithmetic A', p_1 (or d_1) has no place, so $d_{\sim j}$ in S' effectively means d_2 through d_j.

Let the maximum possible odd R-sect generated by primes multisectors $d_{\sim j}$ be denoted by $\mathrm{m}R(d_{\sim j})$, and the maximum possible odd Q-sect generated by prime multisectors $d_{\sim j}$ be denoted by $\mathrm{m}Q(d_{\sim j})$. Here, Spencer-Brown reuses the notation $\mathrm{m}R(d_{\sim j})$ and $\mathrm{m}Q(d_{\sim j})$ to denote the *lengths* of these maximum possible R-sects and Q-sects, respectively. Hence, Spencer-Brown incorporates both recipes in Lemma 3 as follows:

Lemma 3:
The maximum possible odd R-sect $\mathrm{m}R(d_{\sim j})$ of consecutive elements in S' that can be struck using all the odd prime multisectors d_2 through d_j is
(1) $\mathrm{m}R(d_{\sim j}) = d_{j-1} - 1$,
and this sect, counting mirror images as identical, is unique. ...
Alternatively, the maximum odd Q-sect will take us to the brink of the next greater prime than p_j, yielding the formula
(2) $\mathrm{m}Q(d_{\sim j}) = \frac{d_{j+1}-1}{2} - 1$.

The observation that $\mathrm{m}R(d_{\sim j}) = d_{j-1} - 1$ was first made by Legendre in *Essai sur la Théorie des Nombres*, Quatrième Partie, Section IX: "Démonstration de divers théorèmes sur les progressions arithmétiques," pp. 399-406. In addition, Spencer-Brown remarks: "In fact Lemma 3 identifies the maximum possible sect for the pg of every n^2 except in case $n = 23$, for which (uniquely) there are six maximum odd sects of 19". See Appendix 3 for the six unique maximum odd sects generated by prime multisectors $d_{\sim 9}$. A stronger version of Lemma 3 states "no odd sect can exceed the maximum possible odd Q-sect or R-sect by more than 2".

Spencer-Brown mentions that proving this stronger version of Lemma 3 is unnecessary. He claims that upon proving Lemma 4, it will become evident that, for $n > 5$, the maximum sect that can appear *in the segment* must be increasingly shorter than the maximum sect that can appear *in the quantum*. This assertion is under investigation. The following page illustrates Spencer-Brown's proof of Lemma 3. See section 5.5 of this paper, where we review this algorithm from ground up.

Proof of Lemma 3:

Suppose we have found Lemma 3 to be true (as it is) for all odd decimators $d_{\sim j}$ when $j = 4$, and for all lesser values of j. We have

$$\begin{array}{cccc} \bullet & \bullet & \bullet & \bullet \\ 3 & 5 & 7 & 3 \end{array}$$

This is the last case where the R-sect and the Q-sect are identical. As a Q-sect, it is merely the strike of odd prime divisors on the positive side of zero, notably over the numbers 3 through 9. As an R-sect, we imagine both 3's in their paradigmatic places on either side of zero, and arbitrarily choose 5, 7 to fill the two holes in the strike left by the improper divisors 1 and −1. Now consider an induction on j. To introduce the next decimator, in this case, $d_5 = 11$, we take $d_3 = 5$ from its nonparadigmatic place between the 3's and replace it paradigmatically at both ends of the strike, using d_5 to fill the hole in the middle left by d_3. In each case we complete the strike, if we can, with the smaller d's already in use. Thus

$$\begin{array}{cccc} \bullet & \bullet & \bullet & \bullet \\ 3 & 5 & 7 & 3 \end{array}$$

is followed by

$$\begin{array}{cccccc} \bullet & \bullet & \bullet & \bullet & \bullet & \bullet \\ 5 & 3 & 11 & 7 & 3 & 5 \end{array}$$

is followed by

$$\begin{array}{cccccccccc} \bullet & \bullet & \bullet & \bullet & \bullet & \bullet & \bullet & \bullet & \bullet & \bullet \\ 3 & 7 & 5 & 3 & 11 & 13 & 3 & 5 & 7 & 3 \end{array}$$

and so on.

We say a multisector is in a *paradigmatic place*, or pp, if it imbeds in a point that can be identified with the associated prime in A'. Otherwise we say it is in a nonparadigmatic place np. In any R-sect, the two central multisectors begin by being np. Then, by introducing the next larger multisector, the smallest of them goes from np to pp.

Because at every addition of 1 to j there is no alternative to what we may do, this validates the induction and proves the uniqueness of the result. QED

Lemma 4: (4) $d_{j-1} - 1 \geq \frac{d_{j+1}-1}{2} - 1$.

In other words, $\mathrm{m}R(d_{\sim j}) \geq \mathrm{m}Q(d_{\sim j})$.

Note that Spencer-Brown does not give a standard arithmetical proof of Lemma 4 but gives a proof of Lemma 4 by *elementals*, which we outline in section 5 of this paper.

The stronger version of Lemma 4 is $\mathrm{m}R(d_{\sim j}) > \mathrm{m}Q(d_{\sim j})$, for $j > 4$, that is, the maximum R-sect is *strictly greater than* the maximum Q-sect for $j > 4$.

Rearranging the terms in Lemma 4 yields a powerful relationship between the "previous" prime p_{j-1} and the "successive" prime p_{j+1}. Note, $p_j = d_j$.

$$p_{j+1} \leq 2p_{j-1} + 1.$$

An illustration of the above statement for small primes is as follows:
$5 \leq 2(2) + 1$.
$7 \leq 2(3) + 1$.
$11 \leq 2(5) + 1$.
$13 < 2(7) + 1$. Note: From this point onward, the relationship is a strict inequality.
$17 < 2(11) + 1$,
...

The reader is encouraged to verify $p_{j+1} \leq 2p_{j-1} + 1$ for larger primes!

What is profound is that Lemma 4 controls how much bigger the $(j + 1)^{\text{th}}$ prime is in relation to the $(j - 1)^{\text{th}}$ prime. In other words, there is a feedback from the jth prime to the $(j - 1)^{\text{th}}$ prime and a feed-forward to the $(j + 1)^{\text{th}}$ prime.

Following the injunctions provided by Legendre in Part 4, Section 9 of *Essai sur la Théorie des Nombres* and Spencer-Brown in his proof of Lemma 3, Appendix 8, *Laws Of Form*, we construct what we call the "sect pyramid" (or the Spencer-Brown sieve or GSB sieve). When viewing this pyramid, compare the maximum odd R-sect of line L with the maximum odd Q-sect of line L+2 to see Lemma 4 in action.

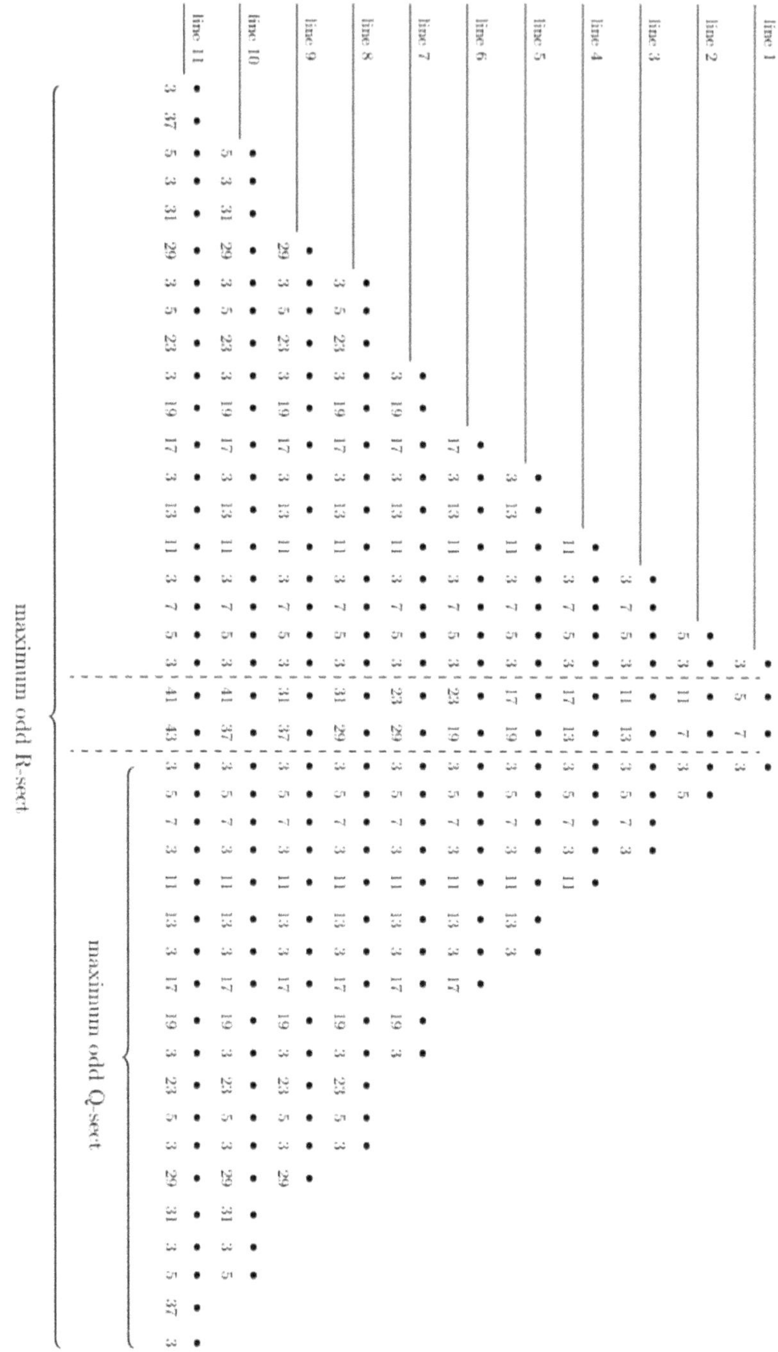
line 1
line 2
line 3
line 4
line 5
line 6
line 7
line 8
line 9
line 10
line 11
maximum odd R-sect
maximum odd Q-sect

3.3 Quadrantic and Reflexive sects

The maximum odd Q-sect m$Q(d_{\sim j})$ can be constructed by marking the odd numbers between 1 and p_{j+1} and replacing each odd composite by its least prime factor.

Consider m$Q(d_{\sim 3})$. Marking the odd numbers upto (but not including) $p_4 = 7$, the maximum odd Q-sect m$Q(p_{\sim 3})$ is $\underset{3}{\bullet}\;\underset{5}{\bullet}$ (and its mirror image $\underset{5}{\bullet}\;\underset{3}{\bullet}$).

Consider m$Q(d_{\sim 4})$. Marking the odd numbers upto $p_5 = 11$ (and replacing each odd composite by its least prime factor), the maximum odd Q-sect m$Q(d_{\sim 4})$ is $\underset{3}{\bullet}\;\underset{5}{\bullet}\;\underset{7}{\bullet}\;\underset{3}{\bullet}$ (and its mirror image $\underset{3}{\bullet}\;\underset{7}{\bullet}\;\underset{5}{\bullet}\;\underset{3}{\bullet}$).

In general, it is important to note that the maximum odd Q-sect m$Q(d_{\sim j})$ appears on the number line mirrored about p_j-primorial, i.e. at $p_j\# = p_1 \times p_2 \times \dots p_j$ and integral multiples of $p_j\#$. This is evident when studying the prime paradigm or full reflexive strike as shown by Spencer-Brown in Figure 1 of Appendix 8.

To illustrate the idea that the maximum odd Q-sect m$Q(d_{\sim j})$ appears on the number line mirrored about $p_j\#$, consider m$Q(d_{\sim 5})$. Using prime multisectors d_2 through d_5, the maximum odd Q-sect, m$Q(p_{\sim 5})$, is $\underset{3}{\bullet}\;\underset{5}{\bullet}\;\underset{7}{\bullet}\;\underset{3}{\bullet}\;\underset{11}{\bullet}$ (and its mirror image $\underset{11}{\bullet}\;\underset{3}{\bullet}\;\underset{7}{\bullet}\;\underset{5}{\bullet}\;\underset{3}{\bullet}$). Note that p_5-strike or $p_5\# = 11\# = 2 \times 3 \times 5 \times 7 \times 11 = 2310$. On the number line, observe that m$Q(d_{\sim 5})$ is mirrored about $p_5\#$, which is like an origin:

|	1	3	5	7	9	11	13	
|	•	•	•	•	•	•	•	…
|		3	5	7	3	11		

2297	2299	2301	2303	2305	2307	2309	|	2311	2313	2315	2317	2319	2321	2323
•	•	•	•	•	•	•	|	•	•	•	•	•	•	•
	11	3	7	5	3		|		3	5	7	3	11	
							11#							

On the other hand, the maximum odd R-sect m$R(d_{\sim j})$ is constructed by first placing

p_j and p_{j-1} about a "mirror" (call this a *local* mirror), then appending the maximum odd Q-sect m$Q(p_{\sim(j-2)})$ and its mirror image on either side of this local mirror.

For example, m$R(d_{\sim 5})$ is constructed by placing $p_5 = 11$ and $p_4 = 7$ about a local or internal "mirror", then appending m$Q(d_{\sim 3})$ and its mirror image to either side:

•	•	•	|	•	•	•
5	3	7	|	11	3	5

or

•	•	•	|	•	•	•
5	3	11	|	7	3	5

Since the R-sect m$R(d_{\sim j})$ is composed of two mirrored Q-sects m$Q(p_{\sim(j-2)})$, the local (or internal) mirror must be located at $p_{j-2}\#$ or some integral multiple of $p_{j-2}\#$. For the example of m$R(d_{\sim 5})$, the local (or internal) mirror is at $4 \cdot (5\#)$ and $73 \cdot (5\#)$.

The maximum odd R-sect m$R(p_{\sim 5})$ and its mirror image appear on the number line mirrored about $\frac{p_5\#}{2} = \frac{11\#}{2} = 1155$, which we will call the *global* mirror or half-mirror. This demarcates the first quadrant or the *quantum* (See Figure 1).

			$4 \cdot (5\#)$									$73 \cdot (5\#)$			
115	117	119	|	121	123	125		|		2185	2187	2189	|	2191	2193	2195
•	•	•	|	•	•	•	…	|	…	•	•	•	|	•	•	•
5	3	7	|	11	3	5		$\frac{11\#}{2}$		5	3	11	|	7	3	5

Next, m$R(d_{\sim 6})$ is constructed by placing $p_6 = 13$ and $p_5 = 11$ about a local or internal "mirror" and then appending m$Q(d_{\sim 4})$ on either side. We know that for m$R(d_{\sim 6})$, the local mirror will be at some integral multiple of $p_4\#$. Using a simple code in Python, we are able to find the first appearance of m$R(d_{\sim 6})$ centered at $45 \cdot (7\#)$:

					$45 \cdot (7\#)$					
9441	9443	9445	9447	9449	|	9451	9453	9455	9457	9459
•	•	•	•	•	|	•	•	•	•	•
3	7	5	3	11	|	13	3	5	7	3

The global mirror is at $\frac{p_j\#}{2} = \frac{13\#}{2} = 15015$, so the mirror image of this R-sect is symmetrically placed on the other side of this global mirror, centered at $98 \cdot (7\#)$:

$98 \cdot (7\#)$

20571	20573	20575	20577	20579	20581	20583	20585	20587	20589
•	•	•	•	•	•	•	•	•	•
3	7	5	3	13	11	3	5	7	3

3.4 A short proof of Theorem 1

The following proof is slightly different from the proof provided by Spencer-Brown. This proof assumes Lemmas 1A, 2, 3, and 4:

Theorem 1:
There is at least one odd prime between n^2 and $(n+1)^2$.

Proof:
There are n odd numbers between n^2 and $(n+1)^2$ (Lemma 2).

The least prime factor of any odd composite between n^2 and $(n+1)^2$ is $\leq n$ (Lemma 1A). Let $p_1, p_2, p_3, \dots, p_j$ be all the primes $\leq n$.

The longest consecutive stretch of odd composites using $p_2, p_3, \dots, p_j$ is either m$R(d_{\sim j})$ of length $p_{j-1} - 1$, or m$Q(d_{\sim j})$ of length $\frac{p_{j+1}-1}{2} - 1$ (Lemma 3).

For $j > 4$, we know that the length of m$R(d_{\sim j})$ > the length of m$Q(d_{\sim j})$ (Lemma 4).

To prove Theorem 1, the maximum stretch of odd composites between n^2 and $(n+1)^2$ cannot be greater than m$R(d_{\sim j}) = p_{j-1} - 1$. Now $p_{j-1} - 1 < n$, since $p_{j-1} < n$.

The consecutive odd composites between n^2 and $(n+1)^2$ can only factor into primes up to p_j, so the longest length of consecutive odd composites must be $\leq p_{j-1} - 1$, which is $< n$. There are n odd numbers between n^2 and $(n+1)^2$ and the longest consecutive stretch of odd composites in this interval is $< n$. Therefore, one of these n odd numbers must be prime. □

4 Generalizing Theorem 1

Spencer-Brown's proof of Theorem 1 is subtly different from the proof in section 3.4, but both of these proofs rest on the validity of Lemmas 3 and 4. We will see Spencer-Brown's proof in section 5.1. To prove Lemma 4, Spencer-Brown first generalizes Theorem 1 to Theorem 1A, and proceeds to make powerful generalizations.

4.1 The first ratchet

Theorem 1A:
If x is a real number ≥ 1, the next prime $> x$ is at a maximum distance of $2\sqrt{x}$ from x.

In other words, Theorem 1A implies that $p_{j+1} \leq p_j + 2\sqrt{p_j}$. This profound statement offers a sharp bound on gaps between primes and it can be sharpened as the primes get larger. This sharpening is called the *ratchet phenomenon.*

Proposition 1:
Theorem 1A implies Theorem 1.
Proof: Consider the numbers between n^2 and $(n+1)^2$. Let p_j be the largest prime number so that $p_j < n^2$. Let p_{j+1} be the next prime number after p_j. Then, by Theorem 1A, $p_{j+1} \leq p_j + 2\sqrt{p_j}$. We have $p_j < n^2$. Therefore $\sqrt{p_j} < n$, whence $p_{j+1} < n^2 + 2n$. Since $(n+1)^2 = n^2 + 2n + 1$, this implies that $p_{j+1} < (n+1)^2$. Hence, there exists a prime p_{j+1} in between n^2 and $(n+1)^2$. □

Proposition 2 (Andrica equivalent):
$p_{j+1} \leq p_j + 2\sqrt{p_j}$ implies Andrica's conjecture[5] $\sqrt{p_{j+1}} - \sqrt{p_j} < 1$.
Proof: Consider, $p_{j+1} \leq p_j + 2\sqrt{p_j}$ implies $p_{j+1} < p_j + 2\sqrt{p_j} + 1$. It follows that $\sqrt{p_{j+1}} < \sqrt{p_j} + 1$. Hence, $\sqrt{p_{j+1}} - \sqrt{p_j} < 1$. □

[5]Andrica, D. "Note on a conjecture in prime number theory". *Studia Univ. Babeş-Bolyai Math. 1986.* **31** (4): 44-48.

4.2 Ratchet values and ratchet primes

Spencer-Brown recasts Theorem 1A in an even more general form:

(5) $p_{j+1} \leq p_j + \frac{1}{k}\sqrt{p_j} \qquad (k = \frac{1}{2})$,

Call k a *ratchet value*. In Theorem 1A, $k = \frac{1}{2}$.

Now consider, if $p_{j+1} = p_j + \frac{1}{k}\sqrt{p_j}$. This implies $\frac{1}{k}\sqrt{p_j} = p_{j+1} - p_j$. Hence, $\frac{1}{k} = \frac{p_{j+1}-p_j}{\sqrt{p_j}}$ or $k = \frac{\sqrt{p_j}}{p_{j+1}-p_j}$. With this formula in place, one can try values of k for various choices of j. Call *ratchet points* (alternatively, *ratchet primes*) those primes p_j such that for all $l > j$, $p_{l+1} \leq p_l + \frac{1}{k}\sqrt{p_l}$.

The first few ratchet primes are provided by Spencer-Brown as $7, 113, 1327, 1669, \ldots$. The corresponding ratchet values $k = \frac{\sqrt{p_j}}{p_{j+1}-p_j}$ are $k_1 = \frac{\sqrt{7}}{4} = 0.661438...$, then $k_2 = \frac{\sqrt{113}}{14} = 0.7592961...$, then $k_3 = \frac{\sqrt{1327}}{34} = 1.0714120....$

> We can substitute $k = \frac{\sqrt{7}}{4} = 0.661438...$ when (5) becomes an equality for $i = 4$ and remains unequal for all other values of i. This depends on knowing that the jump from 7 to 11 is the biggest jump in the whole prime paradigm, when compared with the size of the take-off point.
> As soon as this jump has been achieved, k advances to the next ratchet point, which occurs at the jump of 14 integers from 113 to 127 making $k = \frac{\sqrt{113}}{14} = 0.7592961...$ A ratchet point is where k advances to a new value from which it cannot slip back. After 7, we know that 113 is the next ratchet point because the jump via 114 to 126 is over a maximum R-sect including 11^2. When this jump has been achieved, k advances to a value > 1, the next ratchet point being $\frac{\sqrt{1327}}{34} = 1.0714120....$

In the handwritten manuscript of *Primes Between Squares*, Spencer-Brown writes:

> It is fairly easy to find the ratchet points by empirical observation but to prove they are ratchet points is more difficult. They occur at record jumps between primes, but of course not every record jump indicates a ratchet point. If that were so, the mathematics would be much too easy.

There is no known formula for the "next" ratchet point, but entire tables can be made. In the Online Encyclopedia of Integer Sequences (OEIS), there is a sequence of primes associated with Andrica's conjecture produced by Harry J. Smith (OEIS A084974: $7, 113, 1327, 1669, 2477, \ldots$). It turns out that Smith's prime sequence is apparently identical with the sequence of ratchet primes for our formulation.

To see how Spencer-Brown's ratchets work in relation to Andrica's conjecture, consider $\sqrt{p_{j+1}} - \sqrt{p_j} < \epsilon$ and examine how large the primes need to be to satisfy this inequality for a given ϵ. If $\sqrt{p_{j+1}} - \sqrt{p_j} < \epsilon$, then $\sqrt{p_{j+1}} < \sqrt{p_j} + \epsilon$. This implies $p_{j+1} < p_j + 2\epsilon\sqrt{p_j} + \epsilon^2$. We can then try ratchet values $\epsilon_j = \sqrt{p_{j+1}} - \sqrt{p_j}$ and look for primes p_j so that for all primes $q > p_j$, the inequality $q' < q + 2\epsilon_j\sqrt{q} + \epsilon_j^2$ holds. Call these special primes p_j *Andrica ratchets.* So far, the Andrica ratchet primes seem to coincide with Spencer-Brown ratchet primes.

The reader is encouraged to see plots on *Mathematica* that show very strongly how $p_{j+1} \leq p_j + \frac{1}{k}\sqrt{p_j}$ is satisfied for ratchet values $k_r = \frac{\sqrt{p_r}}{p_{r+1} - p_r}$, where $p_r = 7, 113, 1327, \ldots$.

Another *Mathematica* experiment is to choose a $k \neq k_r$, that is a value of k that does not correspond to a ratchet prime. The plot behaves chaotically, but then beyond a certain point the inequality $p_{j+1} \leq p_j + \frac{1}{k}\sqrt{p_j}$ holds. This is a numerical illustration of Spencer-Brown's Theorem 1B.

> **Theorem 1B**:
> The size of k can be as large as we please, and $p_{j+1} \leq p_j + \frac{1}{k}\sqrt{p_j}$ will be true of all numbers from some j upwards.

The following is a discrete plot of $p_j + \frac{1}{k}\sqrt{p_j} - p_{j+1}$, for arbitrary $k \approx 2.61008$. Note how from $j \approx 3400$ upwards, $p_j + \frac{1}{k}\sqrt{p_j} - p_{j+1} > 0$.

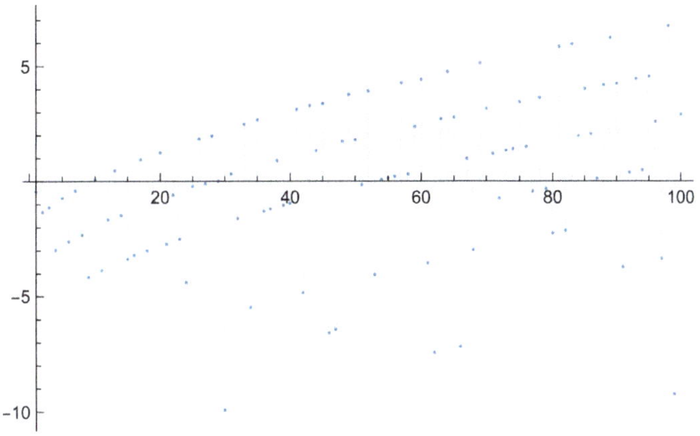
5
-5
-10
20
40
60
80
100

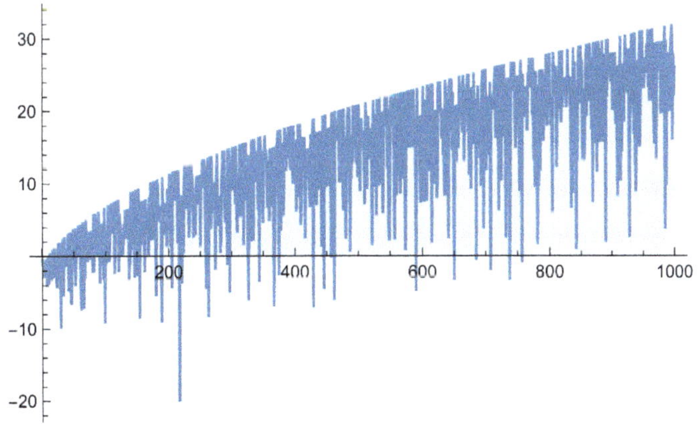
30
20
10
-10
-20
200
400
600
800
1000

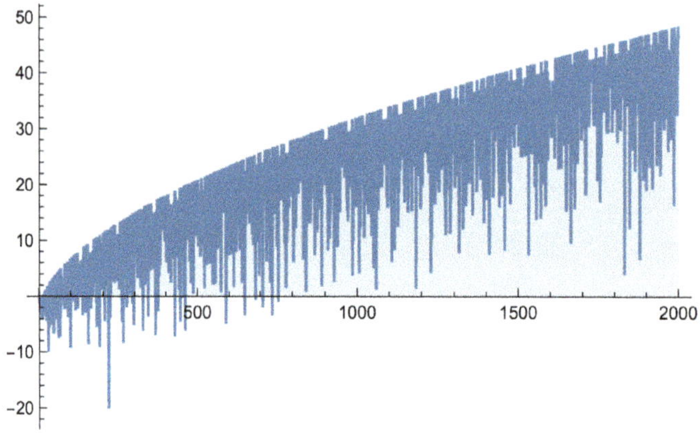
50
40
30
20
10
-10
-20
500
1000
1500
2000

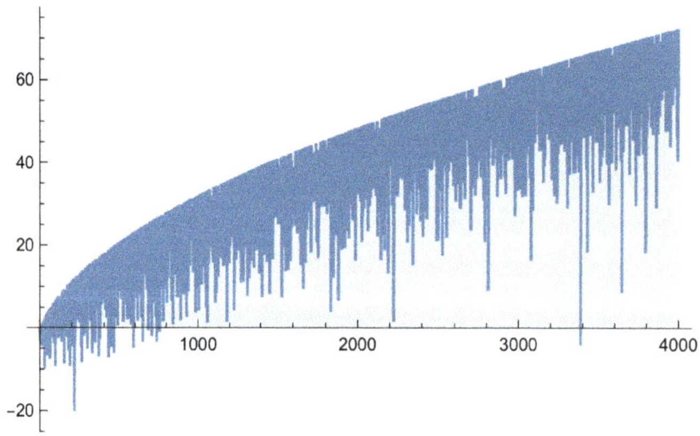
60
40
20
-20
1000
2000
3000
4000

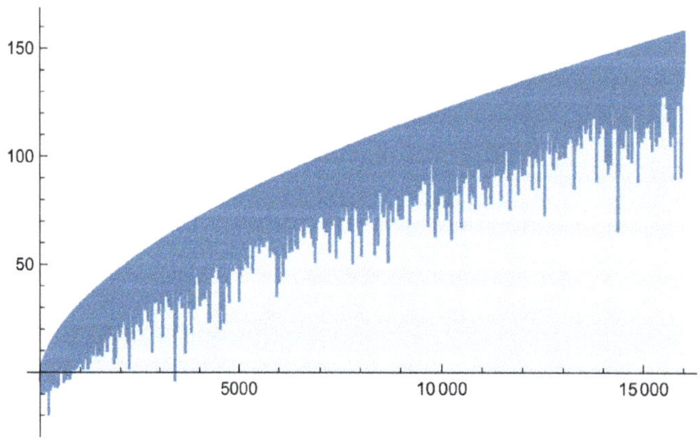
150
100
50
5000
10000
15000

4.3 Haploid and Diploid numbers

Spencer-Brown then provides increasingly powerful and insightful statements about the distribution of primes between squares. The first of these is a corollary,[6] which states that the nearest prime neighbors of any given prime $p_j \geq p_{32}$ lie in the intervals $(p_j - \sqrt{p_j}, p_j)$ and $(p_j, p_j + \sqrt{p_j})$.

Corollary:
If $j \geq 32, (p_{j+1} - p_j)$ and $(p_j - p_{j-1})$ are both less than $\sqrt{p_j}$

Note, the ratchet prime 113 is the 31st prime, p_{31}. This corresponds to the ratchet value of $k = \frac{\sqrt{p_{31}}}{p_{32}-p_{31}} \approx 0.7592961$. The next ratchet prime 1327 is the 217th prime, p_{217}. This corresponds to the ratchet value of 1.0714120. We note that somewhere between p_{31} and p_{217}, the value k crosses a value of 1 and grows larger (Theorem 1B), implying that $\frac{1}{k}$ crosses a value of 1 and grows smaller. Hence, $p_{j+1} \leq p_j + \frac{1}{k}\sqrt{p_j}$ becomes $p_{j+1} \leq p_j + \sqrt{p_j}$ from here on out. This provides some insight into the Corollary.

We also make reference to Spencer-Brown's unpublished "Haploid and Diploid Numbers", in which he writes "The largest haploid number is 127. All numbers greater than 127 are diploid". A haploid number n is a number that has only one prime neighbor in either $(n - \sqrt{n}, n)$ or $(n, n + \sqrt{n})$. A diploid number has a prime neighbor in both $(n-\sqrt{n}, n)$ and $(n, n+\sqrt{n})$. Here, Spencer-Brown is making reference to the Corollary, specifically the ratcheting phenomenon that occurs beyond 127, the 32nd prime, after which $p_{j+1} \leq p_j + \sqrt{p_j}$.

The following is a discrete plot of $k - 1 = \frac{\sqrt{p_j}}{p_{j+1}-p_j} - 1$ on *Mathematica*. Note that the last time $\frac{\sqrt{p_j}}{p_{j+1}-p_j} < 1$ is at $j = 31$. This is a numerical illustration of the Corollary.

[6]See Baker, Harman, Pintz (2000) for a proof of the theorem "There exists x_0 such that for all $x > x_0$, the interval $[x - x^{0.525}, x]$ contains a prime number". In Spencer-Brown's personal copy of A. E. Ingham's *The Distribution of Prime Numbers*, we find the following handwritten conjecture: "p 8 RH ≡ or implied by $p_{n+1} - p_n < p_n^{1/2}$ for some n onwards".

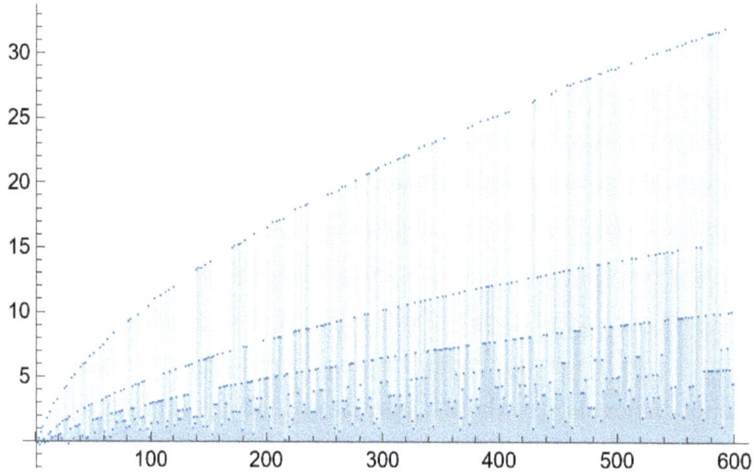

4.4 Half segments

Recall that seg(n) is the stretch of numbers between n^2 and $(n+1)^2$. A *half segment* is the stretch of numbers between n^2 and $n(n+1)$, or between $n(n+1)$ and $(n+1)^2$.

Note that seg(1) is the stretch of numbers between 1 and 4. Halving this segment illustrates the existence of one prime in each half segment, 2 and 3. Similarly, note that seg(2) is the stretch of numbers between 4 and 9. Halving this segment illustrates the existence of one prime in each half segment, 5 and 7. Demonstrate the existence of one prime in each half segment for seg(3) and seg(4). Observe seg(5) below. In this case, the primorial $2 \times 3 \times 5 = 30$ is the vertical line that splits seg(5) into half segments, illustrating the existence of one prime in each half segment, 29 and 31.

5^2	27	29	|	31	33	35	6^2
□	•	•	|	•	•	•	□
5	3		|		3	5	

On page 200 of Appendix 8, Spencer-Brown writes: "I will close with a more-difficult theorem, that I first proved in the summer of 1998". He finally concludes with the following conjecture:

> **Theorem 2**:
> For all natural squares, there is a prime in every half segment[a].
>
> [a] Oppermann's (hitherto unproved) conjecture of 1862 is incomplete, because its wrong notation disallows the listing of 2 between 1^2 and $(1+\frac{1}{2})^2$. See Dickson's *History* Vol 1 p 435. I can thus claim that this delightful theorem is mine.

Ulam's spiral[7] has a variation called the "counterdiagonal of squares" where every completion of a square grid of numbers ends with a square number. This variation clearly demarcates successive square segments as a pair of adjacent edges of the square grid. As seen below, each edge of this grid is a half segment. Theorem 2 states that there exists a prime in every edge of this counterdiagonal of squares:

362	363	364	365	366	367	368	369	370	371	372	373	374	375	376	377	378	379	380	381
361	290	291	292	293	294	295	296	297	298	299	300	301	302	303	304	305	306	307	382
360	**289**	226	227	228	229	230	231	232	233	234	235	236	237	238	239	240	241	308	383
359	288	**225**	170	171	172	173	174	175	176	177	178	179	180	181	182	183	242	309	384
358	287	224	**169**	122	123	124	125	126	127	128	129	130	131	132	133	184	243	310	385
357	286	223	168	**121**	82	83	84	85	86	87	88	89	90	91	134	185	244	311	386
356	285	222	167	120	**81**	50	51	52	53	54	55	56	57	92	135	186	245	312	387
355	284	221	166	119	80	**49**	26	27	28	29	30	31	58	93	136	187	246	313	388
354	283	220	165	118	79	48	**25**	10	11	12	13	32	59	94	137	188	247	314	389
353	282	219	164	117	78	47	24	**9**	2	3	14	33	60	95	138	189	248	315	390
352	281	218	163	116	77	46	23	8	1	**4**	15	34	61	96	139	190	249	316	391
351	280	217	162	115	76	45	22	7	6	5	**16**	35	62	97	140	191	250	317	392
350	279	216	161	114	75	44	21	20	19	18	17	**36**	63	98	141	192	251	318	393
349	278	215	160	113	74	43	42	41	40	39	38	37	**64**	99	142	193	252	319	394
348	277	214	159	112	73	72	71	70	69	68	67	66	65	**100**	143	194	253	320	395
347	276	213	158	111	110	109	108	107	106	105	104	103	102	101	**144**	195	254	321	396
346	275	212	157	156	155	154	153	152	151	150	149	148	147	146	145	**196**	255	322	397
345	274	211	210	209	208	207	206	205	204	203	202	201	200	199	198	197	**256**	323	398
344	273	272	271	270	269	268	267	266	265	264	263	262	261	260	259	258	257	**324**	399
343	342	341	340	339	338	337	336	335	334	333	332	331	330	329	328	327	326	325	**400**

[7] Stein, M. L. Ulam, S. M. Wells, M. B. "A Visual Display of Some Properties of the Distribution of Primes". Amer. Math. Monthly 71, 516-520. 1964.

5 Elementals

5.1 Spencer-Brown's proof of Theorem 1

Spencer-Brown's proof of Theorem 1 in Appendix 8 employs a different form of reasoning. A brief outline of his proof is as follows.

> *(Spencer-Brown's proof of Theorem 1)*
> If Lemma 4 is true, as we are supposing, then Lemma 3 identifies the maximum possible sect that might appear in seg(n), as a result of some strike of all the prime multisectors $\leq n$, and the proof of Theorem 1 can be completed without further difficulty. We must show that the inequality
>
> $$n > \mathrm{m}R(d_{\sim j})$$
>
> is always true when j is the ordinal number of the greatest prime in the pg of n^2.
> When $n = 1$ or $n = 2$, we cannot use just the odd numbers. There is no prime in the pg of 1^2, so all the integers in seg(1), 2 and 3, must be prime, and 3 is odd. There is only one prime 2, in the pg of 2^2, so all the odd numbers, 5 and 7, in seg(2) must be prime. For larger n we can consider the odd numbers only. There can be no odd prime nearer to n on the small side than $n - 2$, and since by Lemma 3, $\mathrm{m}R(d_{\sim j}) = d_{j-1} - 1$, the maximum odd sect must be of length $p_{j-1} - 1$, i.e., is always short of n by at least 3.[a] So the maximum successive strike of odd numbers by the odd primes in the pg of n^2 is at most $n - 3$, so at least one of the n odd numbers between n^2 and $(n+1)^2$ is prime. QED
>
> [a]The special case of $n = 23$ makes to difference, since 23, not being a lesser prime twin, has no effective strike in seg(n).

Note that the proof of Theorem 1 in section 3.4 is similar to the above proof above. Additionally, Spencer-Brown points out a circular relationship between Theorem 1, Lemma 3, and Lemma 4. This is illustrated as follows and described in detail, shortly.

$$\text{Lemma 4} \implies \text{Lemma 3} \implies \text{Theorem 1} \implies \text{Lemma 4} \ldots$$

After his proof of Lemma 3, he writes:

> Furthermore, we are now aware that Lemma 3 will always indicate the maximum[a] possible sect that might appear in seg(n), provided Lemma 4 is true, in other words, provided the primes in the paradigm are not too far apart.
>
> [a]Because for $n > 5$ the pg divisors cannot all be pp in the segment. For $n > 5$ the maximum sect in the segment will be shorter by an increasing margin than the maximum R-sect, provided the inequality in Lemma 4 is true by an increasing margin, as I shall prove it must be.

According to Spencer-Brown in the above excerpt, Lemma 3 is contingent on Lemma 4. And in subsequent excerpts, he states that Lemma 4 is contingent on Theorem 1. To prove Lemma 4, Spencer-Brown constructs Theorem 1A by "substituting a continuous variable x for the stochastic variable n in Theorem 1" and introduces fake prime tables, Table 1 and Table 2.

Starting with the idea of "not too far apart", Spencer-Brown considers the extreme case that the primes $p_j > 5$ are distributed in such a way that the next prime, p_{j+1}, would be of the form $2p_j - 3$. This is evidently not how the primes are distributed in reality but this experiment is used by Spencer-Brown to show that Bertrand's postulate (Tchebychef's theorem[8]) is "misleadingly weak" and the resulting calculations for maximum R and Q sects cannot be possible. On page 195 of Appendix 8, he writes:

> Suppose we could show that Theorem 1A is true in all such instances of x after a certain point. How would this affect Lemma 4? First we construct a table of maximum distances between successive primes that would render Lemma 4 barely true.

Table 1 of Appendix 8 is a fake prime table. Here, Spencer-Brown follows a line of reasoning similar to his earlier assumption of Bertrand's postulate, now assuming that Lemma 4 is "barely true". Lemma 4 states that $\mathrm{m}R(d_{\sim j}) \geq \mathrm{m}Q(d_{\sim j})$. By Lemma 4 being "barely true", Spencer-Brown means $\mathrm{m}R(d_{\sim j}) = \mathrm{m}Q(d_{\sim j})$, i.e., he assumes that $p_{j+1} = 2p_{j-1} + 1$. Using this relationship, he determines the next "prime", as shown above, in column 2.

[8]For $n > 3$, there is a prime between n and $2n - 2$.

n of "p_n"	next "prime"	difference
0	[1]	
		1
1	2	
		1
2	3	
		2
3	5	
		2
4	7	
		4
5	11	
		4
6	15	
		8
7	23	
		8
8	31	
		16
9	47	
		16
10	63	

Table 1. Maximum distances between "primes" compatible with Lemma 4 being barely true. If n is even, the nth "prime" must be $2^{n/2+1}-1$, and if n is odd, the nth "prime" must be the mean of its neighbours.

Table 2 of Appendix 8 is also a fake prime table.[9] Spencer-Brown defines "successive" "prime" here as the nearest odd number within the bare limits of Theorem 1A. In other words, a "successive" "prime" is the nearest odd number within two square-roots of the actual prime p_{j-1} in column 1. To read this table, start with p_{j-1}, which is the real prime. Take 11, for example, then the next prime is $11+2\sqrt{11} \approx 17$, and the next prime is $17+2\sqrt{17} \approx 25$.

For "successive" "primes"			actual maximum	theoretial maximum	
p_{j-1}	p_j	p_{j+1}	R-sect	Q-sect	difference
11	17	25	10	11	−1
13	19	27	12	12	0
17	25	35	16	16	0
19	27	37	18	17	+1
23	31	41	22	19	+3
3001	3109	3219	3000	1608	1392
1000003	1002003	1004005	1000002	502001	498001

Table 2. Comparisons of maximum R-sects and maximum Q-sects for "successive" "primes" barely within the limits of Theorem 1A. While the R-sect remains at $p_{i-1}-1$, the Q-sect approaches a limit of half this value. The p_{i-1} are real primes, but the p_i and p_{i+1} are the nearest odd numbers within the limits set by the theorem.

[9]Note: The caption of Table 2 on page 196 of *Laws Of Form* (Bohmeier Verlag, 2015) should be corrected to p_{j-1}, p_j, and p_{j+1}.

Using Table 1 and Table 2, Spencer-Brown provides the following analysis on page 197:

> Comparing where the tables become comparable, we see that, for $p_{j-1} = 23$, Table 2 makes a span 23, 31, 41, while Table 1 makes a wider span 23, 31, 47, so Table 2 from here on will always give values within the limits of Lemma 4 being true. In fact it begins to happen at $p_{j-1} = 13$, and from $p_{j-1} = 19$ the maximum R-sect is always greater than the maximum Q-sect. Thus if Theorem 1A is true, it locks Lemma 4 into inequality from $n = 19$ onwards. This in turn ensures that Lemma 3 must indicate the longest possible sect of the multisectors in the pg of n^2 that can appear in seg(n). And if Lemma 3 from then on always indicates the longest possible sect in seg(n) of the multisectors corresponding to the prime divisors in the pg of n^2, then Theorem 1A, and with it Theorem 1, must be true from then on. And if Theorem 1 is true from then on, Lemma 4 is true, and therefore irrelevant, from then on, therefore Lemma 3 indicates the maximum sect in seg(n) from then on, and so on *ad infinitum.*
>
> It is a simple matter to ascertain that Theorem 1 is true for n up to 19, and is true for the intervening values of n to 23, and so on, so that Lemma 4 cannot be false for these values either. So Theorem 1, by proving its own lemma, has rendered itself true for all n, and no other argument is either necessary or relevant.

"In short, Theorem 1, like the root in Newton's formula,[10] hunts out its own truth" (page 198, *Laws Of Form*, Bohmeier Verlag, 2015). In a footnote to this statement, Spencer-Brown writes: "And so do Lemmas 3 and 4, since each elemental of an interlocking set is of equal status. Thus we need no longer consult Table 2 to see experimentally that $\frac{\mathrm{mQ}d_{\sim j}}{\mathrm{mR}d_{\sim j}} \to \frac{1}{2}$ as $j \to \infty$, because the interlocking truth of Lemma 4 with Theorem 1A now assures us *mathematically* that this must be so".

Thus, Spencer-Brown defines *elementals*, a recursive form of mathematical reasoning

[10]Newton's formula or the Newton-Raphson method (named after the method found in Isaac Newton's *Method of Fluxions* and Joseph Raphson's *Analysis Aequationum Universalis*) is an iterative method for finding the roots of an arbitrary function $f(x)$. Starting with an initial guess x_0, a better approximation x_{n+1} is obtained by using the iterative formula $x_{n+1} = x_n - \frac{f(x_n)}{f'(x_n)}$, where $f'(x_n)$ is the derivative of $f(x)$ at x_n. The process is repeated until the approximations become close to the actual root.

whose algebraic representation is a "valid equation of the second degree, with real, i.e., self-confirming, roots".[11] The following two excerpts are from pages 197-198.

> I coined the new noun *elemental* to denote any of a set of consequential theorems that dovetail with each other in this way, so that they all stand or fall together. To prove that any of them is true, it is necessary first to suppose that one of them is, whereafter the resulting truth of the others confirms the truth of the initial one. It is a kind of contrary to an ordinary indirect proof, where we first have to suppose the initial proposition is false, and get a contradiction. Here we can suppose it is true, and get a confirmation.

If one walks through the following argument using *elementals* very carefully, letting the structure count, it has the structure of the modulator function E4, as introduced in Chapter 11 of *Laws Of Form*. We will discuss this in detail shortly.

> I suggest the reader pause for a moment to consider the astonishing nature of this argument. We have three theorems, Lemma 3, considered as indicative of the longest sect that can appear in seg(n), Lemma 4, and Theorem 1. They are all different, i.e., they are not equivalent, and each can vary within certain limits. None of them can be proved unless we assume first that one of them is true. We cannot prove that Lemma 3 produces the maximum sect that can appear in seg(n) unless Lemma 4 is true, we cannot prove Lemma 4 is true unless Theorem 1 is true, and we cannot prove Theorem 1 is true unless Lemma 3 produces the longest sect that can appear in seg(n). All these theorems stand or fall together. If one of them is true, they must all be true, and if one of them is false, they may all be false. Yet there is no way of proving any one of them is true without knowing that another of them is also true.

An illustration of his approach using elementals is provided in the diagram below. Here, each arrow represents implication. It is important to note that this diagram

[11] The quadratic equation $x^2 = ax + b$ can be reformulated as $x = a + \frac{b}{x}$. This can be further reformulated as a continued fraction: $x = a + \frac{b}{a+\frac{b}{a+\frac{b}{a+...}}}$. This is an instance of a valid equation of second degree with self-confirming roots. If the discriminant of this quadratic equation is non-negative, the roots are real numbers. However, if the discriminant of this quadratic equation is negative, the roots are complex numbers. This continued fraction is the root of the quadratic equation.

only captures one chain of implications, and the structure of relationships between these statements, when allowed to count, is seen to follow a modulator relationship.

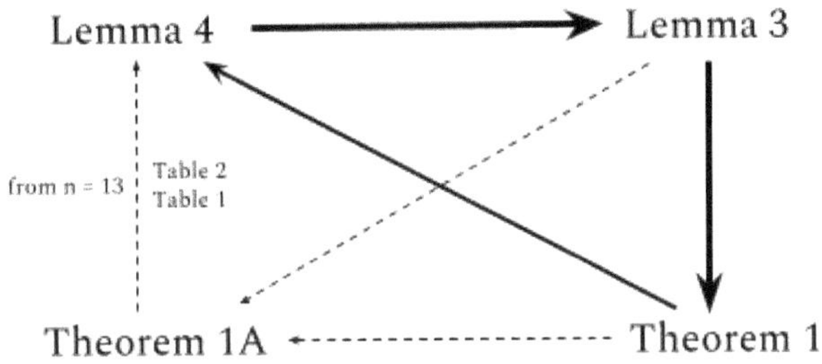

In the "sect pyramid", we can see that the maximum odd R-sect in line 1 and the maximum odd Q-sect in line 3, are equal. This corresponds to the maximum odd sects generated by the first j primes, $j = 4$. From $j = 5$ onward (compare R-sect of line 2 and Q-sect of line 4, and so on, ...), the reflexive sect is greater than the quadrantic sect (Lemma 4) and new holes are introduced. Spencer-Brown uses Table 1 and Table 2 to provide numerical evidence of the first ratchet from $p_4 = 7$ to $p_5 = 11$. The following section, taken from page 200 of Appendix 8 and the handwritten edition of *Primes Between Squares*, points out how beyond this ratchet point, $°\pi$ seg $(n) \geq 3$.

> By associating primes with empty spaces in the strike, we can produce calculated theorems much stronger than Theorem 1. Because $n\#$ grows so much faster than n^2, we see that $n = 5$ is the last time that $n\#$ or any multiple of it can occur within seg(n) with all the pg divisors paradigmatically placed. In this segment we have exactly two primes, 29 and 31, corresponding to the two unoccupied central points in the R-strike of $2, 3, 5$, all of which are pp. From here on at least one of the pg divisors will have to be np, making at least one extra hole in the strike. [Therefore when n is a natural number ≥ 6, $°\pi$ seg $(n) \geq 3$. QED (handwritten)] In practice each np divisor leaves an average of about one extra hole, so an approximate answer for the number of primes in a segment is 2 plus the number of np divisors in the pg.

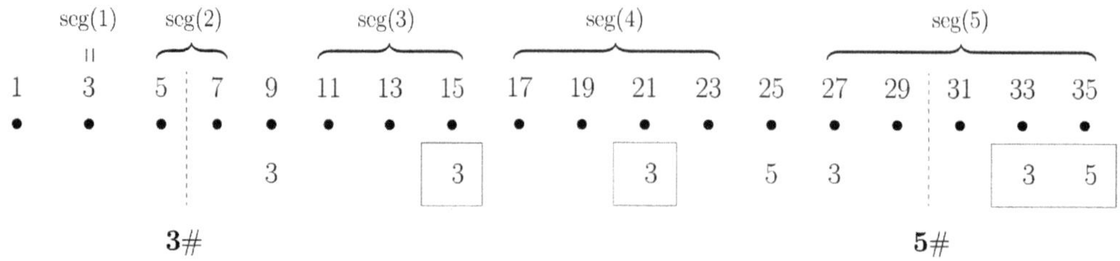

Recall that *n*# refers to the product of all primes up to *n*. In the figure above, observe that 5# is in seg(5) and this is the last time that *n*# can occur within seg(*n*). Note that 7# is in seg(14), and *n*# will always lie outside of seg(*n*) for all *n* > 5. This key excerpt on page 200 of Appendix 8 highlights exactly where the E4 modulator cycle becomes a spiral, starting with the first ratchet. If it were a true modulator, it would repeat itself modulo 4 (resulting in Table 1). The following section reviews the modulator E4 closely in this context.

5.2 Modulators

Chapter 11, *Laws of Form*, is referenced[12] in both printed and handwritten editions of Appendix 8, *Primes Between Squares.* The "modulator" or "reductor" in Chapter 11, called E4, is a re-entering circuit using NOR gates that takes an input 1010 and returns an output 1100 or 0011, i.e., halves the input frequency, over time. In Chapter 11, Spencer-Brown points out that imaginary values are important for the structures of modulators. What Spencer-Brown means by an "imaginary value" is a "transitional state" that maintains a value just long enough to ensure a desired outcome. For modeling purposes, we assume that there is some time delay across the mark but lines transmit information instantaneously. The modulator has stable states that when perturbed by a single bit input make a transition to a new (stable) state. The modulator is designed in such a way that the transition state is unique. With this determinate structure of transitions, the modulator is an entity that can produce a temporal pattern but is itself outside of time, hence, eternal.[13]

[12]"The reader may refer to Appendix 8, where I use an argument of degree 2 to prove that there are primes between squares" (Original Footnote, page 81).

[13]"The reader's attention may be drawn, for example, to the parallel accounts of the emergence of time, i.e. the statements of what we have to do to construct an element that doesn't exist in any of the five orders of eternity" (James Keys, *Only Two Can Play This Game*, pg. 108, 123-132).

There are two ways of understanding modulators—one is by tracking what is happening locally at each marker when the modulator is in a balanced state,[14] and another is by tracking the waveform that emerges globally at each marker. The following are diagrams of four stable states. Each state indicated after the vertical line undergoes a transition to the next one. We do not here go into the details of the transitions.

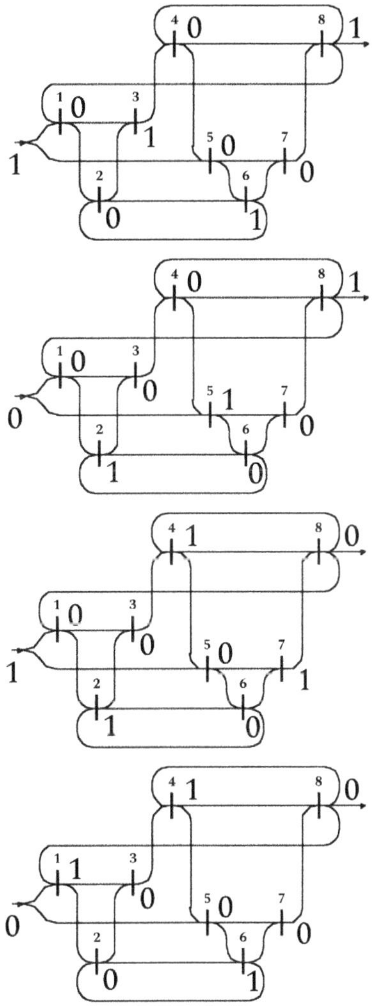

[14]For a lucid discussion on balanced states and transitions of E4, see "Modulators and Imaginary Values" by Kauffman (Chapter 7 of *Laws Of Form—A Fiftieth Anniversary*, 2020).

The waveforms that emerge globally at each marker are as follows:

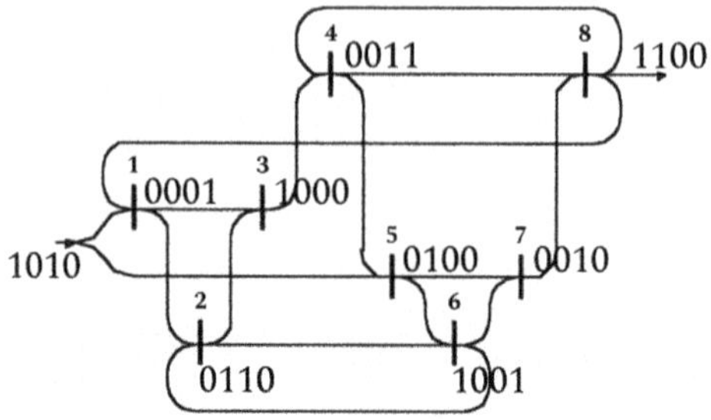

The following table shows both ways of reading the modulator. Reading the table row-wise shows what happens "locally" at each marker at a given balanced state, and reading the table column-wise shows the waveform that emerges "globally" at a given marker. The transition details are not shown in this table.

Input	**1**	**2**	**3**	**4**	**5**	**6**	**7**	**8**	Output
1	0	0	1	0	0	1	0	1	1
0	0	1	0	0	1	0	0	1	1
1	0	1	0	1	0	0	1	0	0
0	1	0	0	1	0	1	0	0	0

We now redraw E4 in a form similar to that conceived by Jeff James.[15] Markers labeled 1-8 are circles labeled 1-8 in this graph. Readers are encouraged to check that this graph is isomorphic to E4:

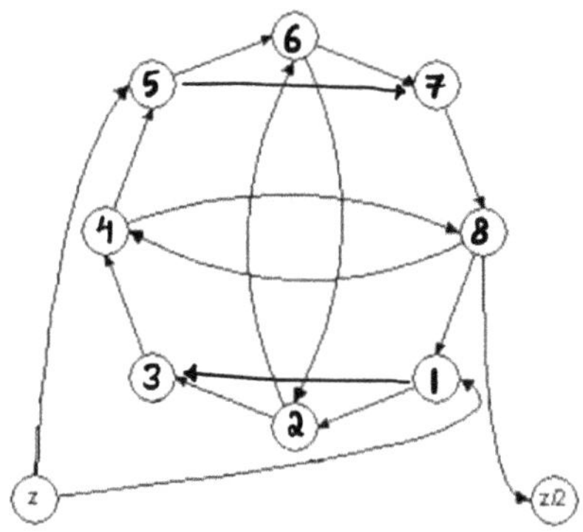

[15]Here, we reference excerpts from "Spencer-Brown's Counter: A simple analysis by Jeff James", a document once archived on the internet, now, lamentably, absent from the archives. Note that Jeff James drew a somewhat different circuit than the one shown above.

While E4 in this form is a sight to marvel at, the local and global ways of looking at E4 in this form are even more astonishing, revealing all its symmetries.[16] First we see what happens locally at each marker at each successive balanced state:

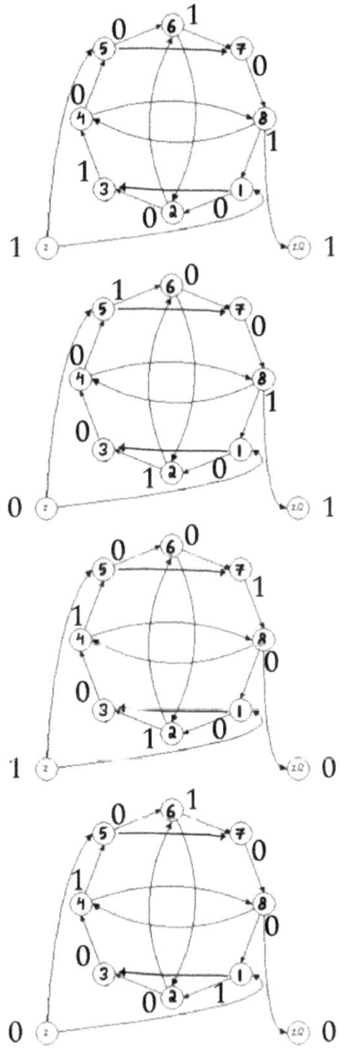

[16]In the words of Jeff James: "The four phases of the counter can be seen below to rotate a triangular constellation of black nodes (*marked 1*) around the circle. The states progress clockwise, as does the constellation within them. Each state represents a settling of the circuit given a new input".

Next, we see the waveforms that emerge globally at each marker:

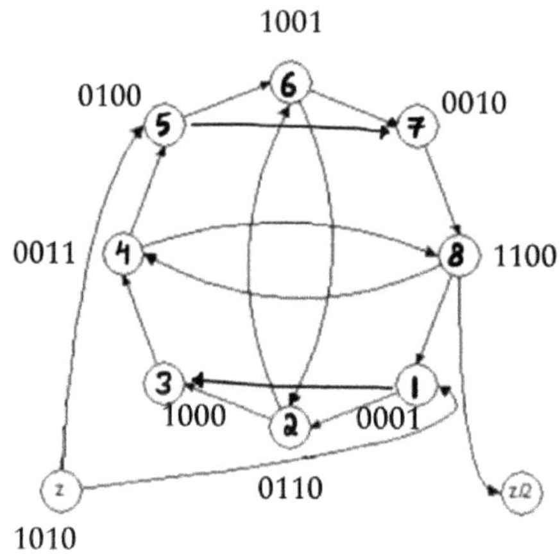

5.3 Binary sieving with E4

We now consider the binary forms of the labels of the markers in E4 and the form of the periodic binary sequences that emerge at these labels. Observe the Jeff James form of E4. First note that the prime markers have single count bitstrings and composite markers have composite unit bitstrings. Consider even markers $2, 4, 6, 8$, and note markers 2 and 6, and markers 4 and 8 are in opposition. Their binary waveforms (marker $4 = 0011$, marker $2 = 0110$, marker $6 = 1001$, marker $8 = 1100$) correspond to numbers 3, 6, 9, and 12, multiples of 3. Now consider markers $1, 3, 5, 7$. Their binary waveforms (marker $1 = 0001$, marker $7 = 0010$, marker $5 = 0100$, marker $3 = 1000$) correspond to $1, 2, 4, 8$, powers of 2. Note, the number 10 corresponds to the input waveform 1010, so the holes created by E4 are the primes $5, 7, 11$. Markers 1 and 5 (corresponding to squares 1 and 4) are in opposition, creating 2 successive spaces that yield the primes 2 and 3. These primes are used as strikers to obtain all primes up to $25 = 5^2$ (Lemma 1A).

Input	**1**	**2**	**3**	**4**	**5**	**6**	**7**	**8**	Output
1	0	0	1	0	0	1	0	1	1
0	0	1	0	0	1	0	0	1	1
1	0	1	0	1	0	0	1	0	0
0	1	0	0	1	0	1	0	0	0

Observe again, the table that illustrates the local (row-wise) and global (column-wise) ways of reading the modulator E4. Reading the modulator E4 row-wise, we see a gap between markers 4 and 5, splitting E4 into two distinct sections. Also note the gap between the second and the third stable state (second row to third row). Observe how opposing quadrants are repeated exactly. Reflexivity balances both these half sections together. Indeed, E4 self-regulates by its own structure.

5.4 Modulator structure in general and binary counting

Note that the general idea of a modulator is that it can accept a waveform or varying input and it will output a waveform of half of the input frequency. Thus if ...10101010... is the input then the output would be ...11001100.... To see how this works, imagine a black box **M** that takes an input x and gives an output y. Write y**M**x for this situation and call it a *stable state* of **M**. It is assumed that once **M** receives x, there will be a time interval after the output y is stably computed. Here are the rules for **M**:

- If y**M**0 is a stable state of **M** and 0 changes to 1, then y**M**1 is a stable state of **M**. In other words, if the input 0 changes to 1, no change happens to the output.
- If y**M**1 is a stable state of **M** and 1 changes to 0, then y'**M**1 is the resulting stable state, where $0' = 1$ and $1' = 0$. In other words, if the input 1 changes to 0, then the output changes (either from 0 to 1 or from 1 to 0).
- One can start the modulator in any of the states 0**M**0, 0**M**1,1**M**0,1**M**1.

Let's see how **M** behaves. Start **M** in the state 0**M**0 and change the input successively, following the rules. We find:

0**M**0
0**M**1
1**M**0
1**M**1
0**M**0
...

As you can see, **M** does indeed behave as advertised. If we put a sequence of modulators together, then the pattern of the outputs corresponds to binary counting:

0M0M0M0
0M0M0M1
0M0M1M0
0M0M1M1
0M1M0M0
0M1M0M1
0M1M1M0
0M1M1M1
1M0M0M0
1M0M0M1
1M0M1M0
1M0M1M1
1M1M0M0
1M1M0M1
1M1M1M0
1M1M1M1

Note that at the next step we go back to 0M0M0M0. This concatenation of three modulators counts to 16. We see that from this, and the fact that the modulator **M** can be built from operations of distinction in the circular pattern of the designs of Chapter 11 (so called "JK-Flip Flop circuits" in the engineering parlance), numerical arithmetic arises from recursive acts of distinction.

5.5 The Spencer-Brown sieve

It is the purpose of this section to point out the general structure of modulators and to point out that this structure is intimately related with the design and behavior of the Spencer-Brown sieve (the GSB sieve). We have discussed this version of the Sieve of Eratosthenes while constructing the "sect pyramid" from Lemma 3 earlier in the paper. Here we start again from the beginning.

Write two place holders $**$. There are two of them. Add one to get 3 and place 3 on either side of the two stars.

$$3 ** 3$$

Now count that to find 4 symbols. Add 1 and get 5. Place 5.

$$53 ** 35$$

Count gives 6, and add one gives 7. Place 7. Note that we are generating prime numbers.

$$753 ** 357$$

But now the 3 can skip to the left by 3 or to the right by 3 and occupy an outer place. When this can be done it must be done.

$$3753 ** 3573$$

One can try to skip 5 by 5 steps, but it will go too far, not at the exact ends. So we are done and we can Count and add one to get 10+1=11, the next prime number.

$$11\ 3753 ** 3573\ 11$$

Neither 3 nor 5 can skip to the ends and so we can count and add one to get $12 + 1 = 13$.

$$13\ 11\ 3753 ** 3573\ 11\ 13$$

Now 3 or 5 can skip forward and strike the ends. When more than one strike can occur, we indicate the smallest one.

$$3\ 13\ 11\ 3753 ** 3573\ 11\ 13\ 3$$

At this point no further strikes are available and so we can Count and add one.

$$17\ 3\ 13\ 11\ 3753 ** 3573\ 11\ 13\ 3\ 17$$

This process continues forever, producing prime numbers. We can think of the GSB sieve as a generalized modulator whose stable states occur when no more strikes are available. At such points the output is the next prime in form of Count +1. Then the modulator can be stablilized again.

Here is what we did in one chart.

$$**$$

$$3 ** 3$$

$$53 ** 35$$

$$3753 ** 3573$$

$$11\ 3753 ** 3573\ 11$$

$$3\ 13\ 11\ 3753 ** 3573\ 11\ 13\ 3$$

$$17\ 3\ 13\ 11\ 3753 ** 3573\ 11\ 13\ 3\ 17$$

$$\ldots$$

The next chart shows how the original modulator pattern is inside this GSB Sieve:

$$* \#$$

$$3 * \#3$$

$$53\# * 35$$

$$3753\# * 3573$$

$$11\ 3753 * \#3573\ 11$$

$$3\ 13\ 11\ 3753 * \#3573\ 11\ 13\ 3$$

$$17\ 3\ 13\ 11\ 3753\# * 3573\ 11\ 13\ 3\ 17$$

$$\ldots$$

One can see a modulator pattern in the center stars. These can be regarded as filled with primes repeating in modulator pattern ready to come out to the boundaries after the strikes have come to stability.

In the following chart, we fill in those central primes. It is on the second appearance of each central prime that it goes outward to take its place in the external row of the GSB Sieve.

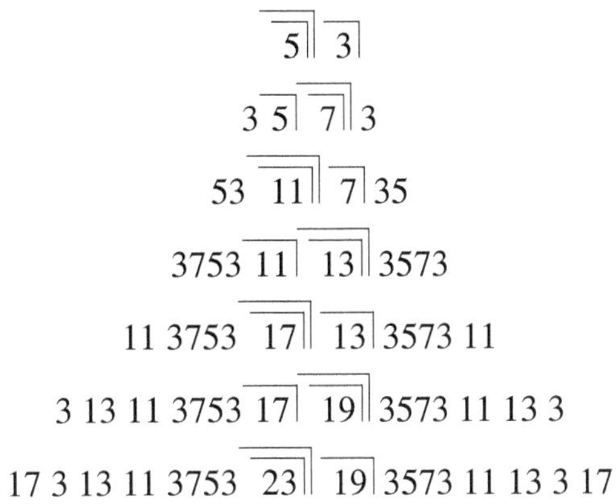

The Spencer-Brown sieve (the GSB sieve) is a powerful evocation of the generation of the prime numbers. Looking at it in this way one can see that it is part of the intricate story of how numbers are generated by circuits of distinctions.

The modulation structure in the central multisectors emerge from two distinct placeholders, i.e., "two successive spaces of expression" (Chapter 11, *Laws Of Form*).

> Every fact about every prime in the universe is contained in our knowledge of the situation of the first prime 2. Furthermore, we do not even need to know that 2 is a prime, merely that it is the first multiplicative integer. 1 is nonexistent as a multiplier, divisor, exponent, or root. Thus 2 can tell us the answer to every question that can possibly be asked about primes.

Following this revelatory excerpt from page 191 of Appendix 8, we close by restating Spencer-Brown's more-difficult theorem, which he "first proved in the summer of 1998".

> **Theorem 2**:
> For all natural squares, there is a prime in every half segment.

Glossary

square segment, or *segment*: The set of whole numbers between successive squares.
seg(n): The set of whole numbers between n^2 and $(n+1)^2$, where $n \in \mathbb{N}$.
seg(x): The set of whole numbers between x^2 and $(x+1)^2$, where $x \in \mathbb{R}$.
°πseg(x): The number of primes between x^2 and $(x+1)^2$.

Theorem 1: There is at least one odd prime in the square segment of every natural number.

Lemma 1: A number h is prime *iff* h is an integer and h does not make an integer quotient with any integer i in case $2 \leq i \leq \sqrt{h}$.

pg of h, or **prime generators of the primes in h:** The set of primes $\leq \sqrt{h}$.

arithmetic A: The system of integers
system S: The mapping of the system of integers onto a line of regularly spaced indistinguishable marks or points stretching endlessly in either direction.

prime multisectors $\mathbf{d_j}$: Operators on the system S that imbed with or strike every second, every third, every fifth, ... every $p_j{}^{\text{th}}$ element, where p_j is the jth prime number. In other words, a prime multisector is all multiples of a given prime.
decimation/decimator: synonymous with multisection/multisector.

A′: The odd integers.
S′: The subsystem of S that maps to the odd integers.

strike:

- (verb) In operating the sieve, we choose a prime number and "strike" out all multiples of that prime number. The operation of the imbedding of a multisector onto the elements of system S: When we speak of imbedding a multisector, this corresponds to striking the corresponding set of points in S.

- (noun) A representative piece of S that results from the striking of a set of multisectors.

- (collective noun) The primorial or prime factorial, also called hash or factorial strike. The primorial of n is the product of primes $\leq n$ and the factorial strike is the strike using these primes. This is also called the full reflexive strike.

$\mathbf{p}_{\sim \mathbf{j}}$: The set of all primes $\leq p_j$.
$\mathbf{d}_{\sim \mathbf{j}}$: The set of all prime multisectors $\leq d_j$.

Figure 1: The **full reflexive strike** (or the **prime paradigm**) of the odd primes $3, 5$. The ringed points are mirror-points at multiples of 2, which do not strictly belong to the odd-strike.

quantum or half-strike: Half of the primorial used to generate the prime paradigm.

Lemma 2: There are exactly n odd numbers between n^2 and $(n+1)^2$.

sect: A consecutive sequence of composite numbers.
R-sect or reflexive sect: A "palindromic" odd sect constructed using Recipe 1.
Q-sect or quadrantic sect: An odd sect constructed using Recipe 2.

Recipe 1: A recipe to create reflexive sects or R-sects. Beginning at a reference mark (0) in a system S, this recipe creates a palindromic odd sect by imbedding or striking every $p_j{}^{\text{th}}$ element (where p_j is the jth prime number) on either side of the reference, including two central points corresponding to -1 and $+1$.
Recipe 2: A recipe to create quadrantic or Q-sects. Beginning at a reference mark (0) in a system S, this recipe creates an odd sect by imbedding or striking every $p_j{}^{\text{th}}$ element (where p_j is the jth prime number) on the right of the reference (zero), including one point corresponding to $+1$.

mR($\mathbf{d}_{\sim \mathbf{j}}$): The maximum R-sect constructed using prime multisectors $d_1, d_2, ...d_j$.
mQ($\mathbf{d}_{\sim \mathbf{j}}$): The maximum Q-sect constructed using prime multisectors $d_1, d_2, ...d_j$.

Lemma 3: The maximum possible sect for the pg of every n^2 is $\mathrm{mR}(d_{\sim j}) = d_{j-1} - 1$ or $\mathrm{mQ}(d_{\sim j}) = \frac{d_{j+1}-1}{2} - 1$. An exception to Lemma 3 is $n = 23$.

Lemma 4: $\mathrm{mR}(d_{\sim j}) \geq \mathrm{mQ}(d_{\sim j})$. In other words, $p_{j-1} - 1 \geq \frac{p_{j+1}-1}{2} - 1$, which when rearranged implies $p_{j+1} \leq 2p_{j-1} + 1$.

paradigmatic place (pp) and nonparadigmatic place (np): A multisector is in a paradigmatic place, or pp, if it imbeds in a point that can be identified with the associated prime in A'. Otherwise it is in a nonparadigmatic place np.

Theorem 1A: If x is a real number ≥ 1, the next prime $> x$ is at a maximum distance of $2\sqrt{x}$ from x.

Table 1: A fake prime table illustrating maximum distances between fake primes corresponding to Lemma 4 being "barely true", i.e., $p_{j+1} = 2p_{j-1} + 1$.

Table 2: A fake prime table that compares actual maximum R-sects (determined by actual primes listed in column 1) with fake maximum Q-sects for fake successive primes that are barely within the limits of Theorem 1A (where a fake successive prime is the nearest odd number within two square-roots of the actual prime in column 1).

elemental: A set of consequential theorems that dovetail with each other in such a way that they stand or fall together.

Proposition (5): $p_{i+1} \leq p_i + \frac{1}{k}\sqrt{p_i} \qquad (k = \frac{1}{2})$.

Ratchet point: A point on the number line where k in Proposition (5) advances to a new value from which it cannot slip back.

Theorem 1B: For any arbitrarily large k, Proposition (5) is true for all numbers from some i upwards.

Corollary: If $i \geq 32, (p_{i+1} - p_i)$ and $(p_i - p_{i-1})$ are both less than $|p_i^{\frac{1}{2}}|$

Theorem 2: For all natural squares, there is a prime in every half segment.

Appendix 1: Sharper Conjectures on Primes Between Squares

Spencer-Brown conjectures that the number of primes between n^2 and $(n+1)^2$ is approximately $\frac{n}{\log n}$. This appendix provides sharper conjectures on the existence of primes between squares as provided by Spencer-Brown.

In Appendix 7, he writes: "The total number $t(n)$ of primes between n^2 and $(n+1)^2$, where n is a natural number greater than 1, is within the limits $A-(B-1) < t(n) < A+(B-1)$ where A is $\frac{n}{\log n}$ and B is $\frac{A}{\log A}$. Since my formula, which I call the Prime Limit Theorem, is not necessarily true unless n is one of the natural numbers $2, 3, 4, ...$, it further confirms the existence of a fundamental relationship between logarithms to the base e and the natural numbers, their squares, and the primes." (pages 182-183, *Laws Of Form*, Bohmeier Verlag, 2015)

"I failed to notice in Appendix 7 of Reference 1 that my upper limit to the possible number of primes in seg(n), $A+(B-1)$ $(A = \frac{n}{\log n}, B = \frac{A}{\log A})$, is also an upper limit to $\pi(n)$." (page 200, *Laws Of Form*, Bohmeier Verlag, 2015)

In *Notes* to Appendix 8, (page 201), Spencer-Brown writes: "The best upper limit to $\pi(x)$ hitherto was Sylvester's 1892 refinement of Tchebychef's result, notably that $0.95695\frac{x}{\log x} < \pi(x) < 1.04423\frac{x}{\log x}$ for all sufficiently large x. We may ignore the lower limit since it is easy to prove that $\frac{n}{\log n} < \pi(n)$ for all $n \geq 17$. Sylvester's upper limit is too small until n is of the order of 10^{11}, whereas my upper limit is true for all n." (pages 201-202, *Laws Of Form*, Bohmeier Verlag, 2015)

The following excerpts are taken from the handwritten version of *Primes Between Squares*, reproduced as follows:

"More importantly my lower limit $A-(B-1) < t$ where A is $\frac{n}{\log n}$, B is $\frac{A}{\log A}$, and t is the total number of primes between n^2 and $(n+1)^2$, provides a much more realistic answer for the minimum number of primes between n and $2n$, and even this can be improved if we make accurate assessments that avoid the approximations due to using logarithms. My upper limit $A+(B-1) > \pi(n)$ is so near to the actual count that a simple modification of it will give an approximation comparable to Riemann's for up to 10^7.

My approximation (one of several I composed in 1998) is $Qx = a + b - c = \pi(x)$ where $a = \frac{x}{\log x}, b = \frac{a}{\log a}, c = \frac{jb}{\log jb}$ and j is an adjustor between 0 and 1. Setting $j = \frac{1}{2}$ gives a very good estimate for $\pi(x)$ in the early reaches of x. My more sophisticated formula (2) in Reference 1 will match Riemann's results as far as we wish to take it, but the summation using a hand calculator is time-consuming for large x. I have since improved it by invoking the subroutine $\mathrm{SB}(x) = \mathrm{Li}\ (x) + \mathrm{Li}\ ((\sqrt{x}-1)^2)/2-1.5$. When x is large we can substitute $\mathrm{SB}([\sqrt{x}+t]^2)$ for the summation term in (2), calling the result $\mathrm{SB}_t(x)$. The difference to the outcome will be negligible. Thus $\mathrm{S}_{1.32}(10^{11}) = 4118052495.21$ and $\mathrm{SB}_{1.32}(10^{11}) = 4118052495.46$ show hardly any difference, except that the former took 12 hours to compute, and the latter only 3 minutes. The Riemann formula for the same x yields $\mathrm{R}(10^{11}) = 4118052494.46$ and is extremely laborious to compute. All the results in Reference 1 were achieved with a programmable hand calculator, using the simplest and most accurate formula for $\mathrm{Li}(x)$: $\mathrm{Li}(x) = \gamma + \log\log x + \sum_{k=1}^{\infty} \frac{\log^k x}{k!k}$.

No one hitherto seems to have remarked that all such approximations to $\pi(x)$ are meaningless unless we state over what stretch of numbers they are claimed to be reasonably accurate. Riemann's and my approximations are claimed to be reasonably accurate near zero, but Li(x) would of course provide a much closer approximation to $\pi(x)$ in the region of a Littlewood crossing.[17] In fact the Riemann-Ramanujan formula[18] is merely a complicated modification of Li(x) to make it a closer approximation to $\pi(x)$ in the region of zero".

[17]See Appendix 9, *Laws of Form* (Bohmeier Verlag, 2015).

[18]Original footnote: The disadvantage of the Riemann-Ramanujan formula is that it requires what is theoretically an infinite number of computations of logarithmic integrals, whereas my subroutine above, whose results converge to those of the R-R formula, requires only two such computations to achieve what is practically the same answer.

Appendix 2: A Constellation of Conjectures

Here, we list a number of well-known conjectures. We will not provide references, since the readers can see these conjectures in a variety of number theory expositions.

Bertrand's postulate:[19] For any integer $n > 3$, there exists at least one prime number p such that $n < p < 2n - 2$.

Legendre's conjecture: For every positive integer n, there is a prime number between n^2 and $(n+1)^2$.

Oppermann's conjecture: For every positive integer n, there is a prime between $n(n-1)$ and n^2, and a prime between n^2 and $n(n+1)$.

Brocard's conjecture: For all n, if p_n is the nth prime number, there are at least four prime numbers between p_n^2 and p_{n+1}^2.

Andrica's conjecture: For all n, if p_n is the nth prime number, $\sqrt{p_{n+1}} - \sqrt{p_n} > 1$.

Cramer's conjecture: For all n, if p_n is the nth prime number, $p_{n+1} - p_n = O((\log p_n)^2)$.

Forgues' conjecture: For all n, if p_n is the nth prime number, $\left(\frac{\log(p_{n+1})}{\log(p_n)}\right)^n < e$.

The following three conjectures are on prime gaps $g_n = p_{n+1} - p_n$:

Firoozbakht's conjecture: For $n \geq 1$, $g_n \leq p_n\left(p_n^{1/n} - 1\right)$.

Nicholson's conjecture: For $n > 4$, $g_n \leq \left((n \ln n)^{1/n} - 1\right)$.

Farhadian's conjecture: For $n > 4$, $g_n \leq p_n\left(\left(p_n \frac{\ln n}{\ln p_n}\right)^{1/n} - 1\right)$.

[19]Although it is called a postulate, it is a theorem that has been proved by Tchebychef and Ramanujan.

The form of Appendix 8 connects Bertrand's postulate, as well as Legendre's, Oppermann's, Brocard's, Andrica's, Firoozbakht's, Cramer's, Farhadian's, Nicholson's, and Forgues' conjectures.

The ratchet points in Theorem 1A and Theorem 1B appear to follow ratchet points in Andrica's conjecture OEIS A084974 "Primes that show the slow decrease in the larger values of the Andrica function $Af(k) = \sqrt{p_{k+1}} - \sqrt{p_k}$, where p_k denotes the k'th prime". Also, see[20] OEIS A182514, A383591, A111943, and A079098.

Denote:
Oppermann's conjecture = O
The standard Andrica conjecture = a or Theorem 1A (Spencer-Brown)
The strong Legendre conjecture = L
The standard Legendre conjecture = l or Theorem 1 (Spencer-Brown).
The strong Brocard conjecture = B
The standard Brocard conjecture = b
Here $x \to y$ means x implies y.

Germán Andrés Paz (2014a) constructs Conjecture 1 such that:
Conjecture 1 $\to l$
Conjecture 1 $\to L$
Conjecture 1 $\to b$
Conjecture 1 $\to O$
Conjecture 1 $\to a$.
Paz (2014b) also finds reentrant reformulations of Conjecture 1.

Matt Visser (2019) constructs a stronger form of the standard Andrica conjecture A and shows $A \to O$. In "Verifying other weaker conjectures", Visser shows $O \to L$, $L \to l$, $L \to B$, $B \to b$, $O \to a$, $a \to l$, and $l \to$ the weak Andrica conjecture.

[20]OEIS A182514 "Primes $prime(n)$ such that $\frac{prime(n+1)^n}{prime(n)} > n$".
OEIS A383591 "Smallest prime p where the absolute difference of the gaps to the adjacent primes exceeds $n\log(p)$".
OEIS A111943 "Prime p with prime gap $q-p$ of n-th record Cramer-Shanks-Granville ratio, where q is smallest prime larger than p and C-S-G ratio is $\frac{(q-p)}{(logp)^2}$".
OEIS A079098 "Conjectured values of greatest k such that for any consecutive primes q, q', $k \leq q < q', \sqrt{q'} - \sqrt{q} < \frac{1}{n}$".

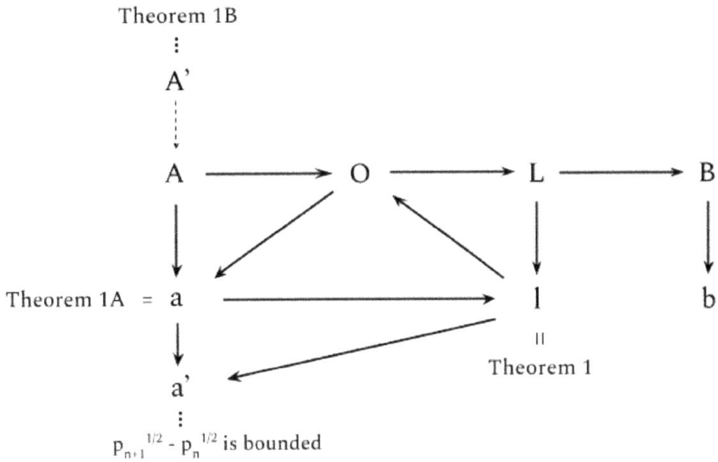

Theorem 1 (Spencer-Brown) is the standard Legendre Conjecture *l*.
Theorem 1A (Spencer-Brown) is the standard Andrica conjecture *a*.

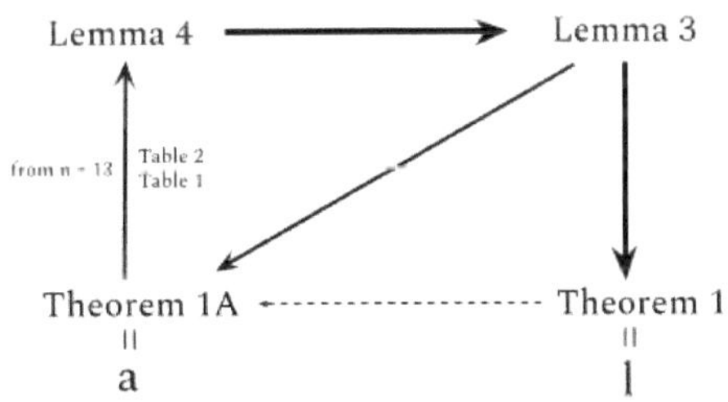

Theorem 1B (Spencer-Brown) pushes the strengthening of Andrica's conjecture to its limit. The following statements are under investigation: Theorem 1B and its Corollary (Spencer-Brown) imply *A* and Conjecture 1.
Theorem 2 (Spencer-Brown) is a strengthening of *O*.

It is fascinating to observe how Theorem 1B, the Corollary, and Theorem 2, relate to the constellation of conjectures on prime gaps.

Appendix 3: Exception at 23

Spencer-Brown's Lemma 3 states that the maximum odd sect is either the R-sect $\mathrm{mR}(d_{\sim j}) = d_{j-1} - 1$ or Q-sect $\mathrm{mQ}(d_{\sim j}) = \frac{(d_{j+1}-1)}{2} - 1$ with a single exception at 23.

For odd primes, $d_{\sim 9}$ = the set of primes d_2 through $d_9 = \{3, 5, 7, 9, 13, 17, 19, 23\}$. Hence, $\mathrm{mR}(d_{\sim 9}) = 19 - 1 = 18$, and $\mathrm{mQ}(d_{\sim 9}) = \frac{(29-1)}{2} - 1 = 13$.

The maximum odd R-sect using $d_{\sim 9}$, written $\mathrm{mR}(d_{\sim 9})$, is of the form:

•	•	•	•	•	•	•	•	•	•	•	•	•	•	•	•	•	•
17	3	13	11	3	7	5	3	23	19	3	5	7	3	11	13	3	17

The maximum odd Q-sect using $d_{\sim 9}$, written $\mathrm{mQ}(d_{\sim 9})$, is of the form:

•	•	•	•	•	•	•	•	•	•	•	•	•
3	5	7	3	11	13	3	17	19	3	23	5	3

However, neither of these is the largest odd-sect using $d_{\sim 9} = \{3, 5, 7, 11, 13, 17, 19, 23\}$. On page 193, Appendix 8, *Laws of Form*, Bohmeier Verlag, Spencer-Brown writes: "In fact Lemma 3 identifies the maximum possible sect for the pg of every n^2 except in case $n = 23$, for which (uniquely) there are six maximum odd sects of 19. ..."

On each of the following two pages, we provide seven maximum odd sects and their mirror images that involve primes $3, 5, 7, 11, 13, 17, 19, 23$. One of these is the maximum odd R-sect $\mathrm{mR}(d_{\sim 9})$ of length 18 as shown above. This can be identified by a dashed line that splits the sect in the middle. Note, this dashed line is located at a multiple of 17#. The other six sects on each of the following two pages are the anomalous maximum odd sects of length 19, one greater than the length stipulated by Lemma 3.

20332473	20332475	20332477	20332479	20332481	20332483	20332485	20332487	20332489	20332491	20332493	20332495	20332497	20332499	20332501	20332503	20332505	20332507	20332509
•	•	•	•	•	•	•	•	•	•	•	•	•	•	•	•	•	•	•
3	5	11	3	13	23	3	7	19	3	17	5	3	11	7	3	5	13	3

24686823	24686825	24686827	24686829	24686831	24686833	24686835	24686837	24686839	24686841	24686843	24686845	24686847	24686849	24686851	24686853	24686855	24686857	24686859
•	•	•	•	•	•	•	•	•	•	•	•	•	•	•	•	•	•	•
3	5	11	3	13	19	3	7	17	3	23	5	3	11	7	3	5	13	3

29609563	29609565	29609567	29609569	29609571	29609573	29609575	29609577	29609579	29609581	29609583	29609585	29609587	29609589	29609591	29609593	29609595	29609597	
•	•	•	•	•	•	•	•	•	•	•	•	•	•	•	•	•	•	
17	3	13	11	3	7	5	3	23	19	3	5	7	3	11	13	3	17	

36068193	36068195	36068197	36068199	36068201	36068203	36068205	36068207	36068209	36068211	36068213	36068215	36068217	36068219	36068221	36068223	36068225	36068227	36068229
•	•	•	•	•	•	•	•	•	•	•	•	•	•	•	•	•	•	•
3	5	11	3	13	17	3	7	23	3	19	5	3	11	7	3	5	13	3

65767863	65767865	65767867	65767869	65767871	65767873	65767875	65767877	65767879	65767881	65767883	65767885	65767887	65767889	65767891	65767893	65767895	65767897	65767899
•	•	•	•	•	•	•	•	•	•	•	•	•	•	•	•	•	•	•
3	5	11	3	13	19	3	7	23	3	17	5	3	11	7	3	5	13	3

82370091	82370093	82370095	82370097	82370099	82370101	82370103	82370105	82370107	82370109	82370111	82370113	82370115	82370117	82370119	82370121	82370123	82370125	82370127
•	•	•	•	•	•	•	•	•	•	•	•	•	•	•	•	•	•	•
3	13	5	3	7	11	3	5	23	3	19	7	3	17	13	3	11	5	3

97689753	97689755	97689757	97689759	97689761	97689763	97689765	97689767	97689769	97689771	97689773	97689775	97689777	97689779	97689781	97689783	97689785	97689787	97689789
•	•	•	•	•	•	•	•	•	•	•	•	•	•	•	•	•	•	•
3	5	11	3	13	23	3	7	17	3	19	5	3	11	7	3	5	13	3

125403081	125403083	125403085	125403087	125403089	125403091	125403093	125403095	125403097	125403099	125403101	125403103	125403105	125403107	125403109	125403111	125403113	125403115	125403117
•	•	•	•	•	•	•	•	•	•	•	•	•	•	•	•	•	•	•
3	13	5	3	7	11	3	5	19	3	17	7	3	23	13	3	11	5	3

140722743	140722745	140722747	140722749	140722751	140722753	140722755	140722757	140722759	140722761	140722763	140722765	140722767	140722769	140722771	140722773	140722775	140722777	140722779
•	•	•	•	•	•	•	•	•	•	•	•	•	•	•	•	•	•	•
3	5	11	3	13	17	3	7	19	3	23	5	3	11	7	3	5	13	3

157324971	157324973	157324975	157324977	157324979	157324981	157324983	157324985	157324987	157324989	157324991	157324993	157324995	157324997	157324999	157325001	157325003	157325005	157325007
•	•	•	•	•	•	•	•	•	•	•	•	•	•	•	•	•	•	•
3	13	5	3	7	11	3	5	17	3	23	7	3	19	13	3	11	5	3

187024641	187024643	187024645	187024647	187024649	187024651	187024653	187024655	187024657	187024659	187024661	187024663	187024665	187024667	187024669	187024671	187024673	187024675	187024677
•	•	•	•	•	•	•	•	•	•	•	•	•	•	•	•	•	•	•
3	13	5	3	7	11	3	5	19	3	23	7	3	17	13	3	11	5	3

193483273	193483275	193483277	193483279	193483281	193483283	193483285	193483287	193483289	193483291	193483293	193483295	193483297	193483299	193483301	193483303	193483305	193483307
•	•	•	•	•	•	•	•	•	•	•	•	•	•	•	•	•	•
17	3	13	11	3	7	5	3	19	23	3	5	7	3	11	13	3	17

198406011	198406013	198406015	198406017	198406019	198406021	198406023	198406025	198406027	198406029	198406031	198406033	198406035	198406037	198406039	198406041	198406043	198406045	198406047
•	•	•	•	•	•	•	•	•	•	•	•	•	•	•	•	•	•	•
3	13	5	3	7	11	3	5	23	3	17	7	3	19	13	3	11	5	3

202760361	202760363	202760365	202760367	202760369	202760371	202760373	202760375	202760377	202760379	202760381	202760383	202760385	202760387	202760389	202760391	202760393	202760395	202760397
•	•	•	•	•	•	•	•	•	•	•	•	•	•	•	•	•	•	•
3	13	5	3	7	11	3	5	17	3	19	7	3	23	13	3	11	5	3

On page 201 of *Laws Of Form* (Bohmeier Verlag, 2015), Spencer-Brown writes: "My theorem that Lemma 3 always denotes a maximum sect has this single exception at $d_{\sim 9}$. The fact never ceases to astonish me, because it means that the primes ≤ 23 can be rearranged to strike more efficiently than they do in the paradigm. It would seem *prima facie* to be impossible, but the fact that it can happen once makes it even more surprising that it can never happen again".[21]

Spencer-Brown also points out the anomaly at 23 in an inequality for Euler's totient function $\phi(n)$, the number of positive integers up to n that are relatively prime to n. This inequality, provided by J.B. Rosser and L. Schoenfeld in "Approximate formulas for some functions of prime numbers" (1962), is as follows:

$$\phi(n) \geq e^{-\gamma} \frac{n}{\log\log n + \frac{5}{2e^{\gamma}\log\log n}}$$

for all $n \geq 3$, with the single exception of $n = 23\#$, for which the constant $\frac{5}{2}$ must be replaced by 2.50637. Here γ is Euler's constant.[22]

Yet another reference to the anomaly at 23 is in the study of finite field extensions of the rational numbers, which is "indispensable to the number theory, even if one's ultimate goal is to understand properties of diophantine expressions and equations in the ordinary integers" (Avigad, 2004). In the Translator's Introduction to "Dedekind's 1871 version of the theory of ideals", Avigad writes: "It can happen, however, that the "integers" in such extensions fail to satisfy unique factorization, a property that is central to reasoning about the ordinary integers. In 1844, Ernst Kummer observed that unique factorization fails for the cyclotomic integers with exponent 23, i.e., the ring $\mathbb{Z}[\zeta]$ of integers in the field $\mathbb{Q}(\zeta)$, where ζ is a primitive twenty-third root of unity" (Avigad, 2014). [23]

That such anomalies exist at 23 is worth the special attention of number theorists.

[21]The anomaly at 23 deserves a paper all on its own (Original footnote, pg. 201, *Laws Of Form*, 2015).

[22]Euler's constant (or the Euler-Mascheroni constant) $\gamma = \int_1^{\infty} \left(-\frac{1}{x} + \frac{1}{[x]}\right) dx \approx 0.57721...$

[23]https://www.andrew.cmu.edu/user/avigad/Papers/ideals71.pdf

Conceived decades before "Primes Between Squares", the 19th and 23rd poems in *23 Degrees of Paradise* written by Spencer-Brown and published under the pseudonym James Keys, anticipate and conjure the deep mathematical insight and mystery of 23.

The 19th poem in *23 Degrees of Paradise* is titled "23".
The 23rd poem is titled "PS":

"*P S*

An idle poet sees and writes
A book of wonders and delights
So full of beauty, life, and love
That poetry's what he writes it of.

Next comes a busy college don
Who looks into the poet's delights
And then a book of learning writes,
And poetry's what he writes it on.

And finally a student, who
Has read the don and learned to scoff,
Cries 'Is this all that poets do?'
Takes his degree and writes it off.

The years pass. A little child,
Rude, untutored, undefiled,
Idle, full of joy, and wild,
Comes to where the Poet Smiled,

And from the vision of it writes
A book of wonders and delights
So full of beauty, life, and love
That poetry's what he writes it of."
(James Keys, *23 Degrees of Paradise*)

References

[1] Andrica, Dorin. 1986. "Note on a Conjecture in Prime Number Theory." *Studia Universitatis Babeş-Bolyai Mathematica* 31, no. 4: 44–48.

[2] Avigad, Jeremy. 2004. "Dedekind's 1871 Version of the Theory of Ideals." *Carnegie Mellon University Repository* (19 March). https://philpapers.org/rec/AVIDV.

[3] Baker, Roger. C., Glynn Harman, and János Pintz. 2001. "The Difference Between Consecutive Primes, II." Proceedings of the London Mathematical Society 83, no. 3: 532–562. https://doi.org/10.1112/plms/83.3.532.

[4] Copi, Irving M. 1971. *The Theory of Logical Types*. London: Routledge & Kegan Paul.

[5] Dickson, Leonard E. 1952. *History of the Theory of Numbers*. Vol. 1: Divisibility and Primality. Washington, DC: Carnegie Institution of Washington.

[6] Ellis, Normandi. 1988. *Awakening Osiris: A New Translation of the Egyptian Book of the Dead*. Grand Rapids, MI: Phanes Press.

[7] Feliksiak, Jan. 2021. "The Nicholson's Conjecture." *ScienceOpen Preprints*, February 24.

[8] Flagg, J. M., Louis H. Kauffman, and Divyamaan Sahoo. 2021. "Laws Of Form and the Riemann Hypothesis." In *Laws of Form: A Fiftieth Anniversary*, edited by Louis H. Kauffman et al. Singapore: World Scientific.

[9] James, Jeffrey. "Spencer-Brown's Counter." https://web.archive.org/web/2010030 6005801/http://www.lawsofform.org/counter/index.html

[10] Kauffman, Louis H. 1977. "Network Synthesis and Varela's Calculus." *International Journal of General Systems* 4: 179–187.

[11] Kauffman, Louis H. 2021. "Modulators and Imaginary Values." In *Laws of Form: A Fiftieth Anniversary*, edited by Louis H. Kauffman et al. Singapore: World Scientific.

[12] Keys, James. 1970. *23 Degrees of Paradise*. W Heffer & Sons Ltd. Cambridge (April 23).

[13] Keys, James. 1972. *Only Two Can Play This Game*. The Julian Press, Inc.

[14] Landau, Edmund. 1912. "Gelöste und ungelöste Probleme aus der Theorie der Primzahlverteilung und der Riemannschen Zetafunktion." *Jahresbericht der Deutschen Mathematiker-Vereinigung* 21: 208–228. http://eudml.org/doc/145337

[15] Legendre, Adrien-Marie. 1808. *Essai sur la Théorie des Nombres.* 2nd ed., Part 4, IX, "Démonstration de divers théorèmes sur les progressions arithmétiques." Paris: Courcier.

[16] Nagura, J. 1952. "On the interval containing at least one prime number". Proc. Japan Acad. 28, no. 4, 177–181.

[17] Online Encyclopedia of Integer Sequences (OEIS). A084974, "Primes that show the slow decrease in the larger values of the Andrica function $Af(k) = \sqrt{p_{k+1}} - \sqrt{p_k}$, where p_k denotes the k-th prime." https://oeis.org/A084974

[18] Panaitopol, Laurenţiu. 2021. "Intervals Containing Prime Numbers." *Notes on Number Theory and Discrete Mathematics* 7, no. 4: 111–114.

[19] Paz, Germán A. 2014a. "On Legendre's, Brocard's, Andrica's, and Oppermann's Conjectures." *arXiv preprint* arXiv:1310.1323 (April 2).

[20] Paz, Germán A. 2014b. "Two statements that are equivalent to a conjecture related to the distribution of prime numbers." arXiv:1406.4801 (June 19).

[21] Petzold, Charles. 2008. *The Annotated Turing: A Guided Tour through Alan Turing's Historic Paper on Computability and the Turing Machine.* Indianapolis: Wiley Publishing, Inc.

[22] Rosser, J. Barkley, and Lowell Schoenfeld. 1962. "Approximate formulas for some functions of prime numbers." *Illinois Journal of Mathematics* 6: 64–94.

[23] Smith, Harry J. 2008. "Andrica's Conjecture."
https://www.oocities.org/hjsmithh/PrimeSR/index.html

[24] Spencer-Brown, George. n.d. "The Universal Operators in Logic." Unpublished manuscript.

[25] Spencer-Brown, George. 1992. "Introduction to Reductors." Unpublished manuscript (December).

[26] Spencer-Brown, George. 1997. *Gesetze der Form.* Hamburg: Bohmeier Verlag.

[27] Spencer-Brown, George. 1998. "Haploid and Diploid Numbers." Unpublished manuscript.

[28] Spencer-Brown, George. 2000. "Primes Between Squares." Handwritten edition.

[29] Spencer-Brown, George. 2007. "Commentary on Appendix 9." Unpublished manuscript.

[30] Spencer-Brown, George. 2015. *Laws of Form.* Revised Sixth English edition. Bohmeier Verlag.

[31] Spencer-Brown, George. 2020. *Laws of Form.* Revised Seventh English edition. Bohmeier Verlag.

[32] Stein, Myron L., Stanisław M. Ulam, and Mark B. Wells. 1964. "A Visual Display of Some Properties of the Distribution of Primes." *American Mathematical Monthly* 71: 516–520.

[33] Visser, Matt. 2019. "Strong Version of Andrica's Conjecture." *International Mathematical Forum* 14, no. 4: 181–188. https://arxiv.org/abs/1812.02762.

www.ingramcontent.com/pod-product-compliance
Ingram Content Group UK Ltd.
Pitfield, Milton Keynes, MK11 3LW, UK
UKHW051935071026
15590UKWH00017B/2083

9 781848 905030